Contents

Truth and Modality for Knowledge Representation

Raymond Turner
University of Essex

Pitman

PITMAN PUBLISHING
128 Long Acre, London WC2E 9AN
A Division of Longman Group UK Limited

© Raymond Turner 1990

First published in Great Britain 1990

British Library Cataloguing in Publication Data

Turner, Raymond, *1947–*
   Truth and modality for knowledge representation.
   1. Philosophy. Representation
   I. Title
   110

   ISBN 0-273-03186-4

ISBN 0 273 03186 4

Printed and bound in Great Britain by
Biddles Ltd, Guildford and King's Lynn

Preface

The need for more expressive systems of knowledge representation is not controversial although it is still debatable whether or not such systems have to be based on formal logic. In this book we shall take it is as read that the formal approach is a worthy one. Our objective is to explore the development of formal languages and appropriate logics for that aspect of knowledge representation concerned with reasoning about truth and modality. A great deal of the current literature in Artificial Intelligence is devoted to the development of formalisms which facilitate the expression of modal concepts. Much of this work, however, is based upon the theories of modality and truth which were developed in the period 1960–1980. In the last ten years there has been a great deal of activity within the logical community centered upon the development of logics of truth and modality. Our objective is to bring this material to the attention of AI researchers by putting it in a context where it might be directly applicable to AI knowledge representation.

Acknowledgements

I must thank Sam Steel and Allan Ramsay for encouraging me to write this book. Nicholas Davies is to be thanked for proof reading the final draft. Roxanna Berretta also proof read the final script. Finally, I cannot find words to convey my thanks to Marisa Bostock and Ann Cook who typed the manuscript.

1 Reasoning Agents

A great deal of research in Artificial Intelligence is concerned with the introduction and exposition of languages of *knowledge representation*. Many of these new languages are variations or extensions of languages which have their origins in the literature of formal logic. The reasons for this are obvious. Formal logic has been concerned throughout its history with the representation of informal argument, its primary goal being the representation of informal arguments in a language where the *form* of the argument and its component sentences are made explicit and unambiguous.

The majority of this formal work in knowledge representation has been carried out within the language and proof theory of the first-order Predicate Calculus. Moreover, many *informal* systems of knowledge representation have been recast within the Predicate Calculus with the implicit belief that such recasting provides a formal semantics for these formalisms and thereby assigns them a degree of clarity and respectability. This may well be true but what is not clear is that the Predicate Calculus is rich enough in expressive power for the full range of applications which knowledge representation seems to subsume. There are many areas where the *natural* representation of informal arguments appears to demand more facilities than those available within the Calculus. The representation of arguments involving time and modality are two topics which immediately spring to mind. Indeed, within the AI community itself (Moore [1980], [1984], Konolige [1986], McDermott [1982], Allen [1984]) such extensions to the Predicate Calculus have been explored with much benefit.

This book is largely concerned with the representation of certain modal notions where *modality* will be given a fairly liberal interpretation. However, before we embark on any lengthy discussion of the nature of modality it seems prudent to examine the general goals of knowledge representation. What exactly is it that we need to represent and reason about?

1.1 Modality and Knowledge Representation

Reasoning agents must be capable of representing and reasoning about what they and other agents *believe, know* and *hold true*. This ability seems to be

1

common ground for many different areas of AI. For example, any natural language program which has to form part of a dialogue system must be capable of representing and reasoning about the knowledge and belief of the players or agents in the system. Moreover, any reasonable system will have to be able to distinguish between these various modalities: to believe a proposition is not the same as knowing it since it is clearly possible for an agent to believe false propositions. A certain tradition in epistemology interprets knowledge as *justified true belief*, and under this interpretation knowing a proposition implies it is both believed and true. On the other hand, it may be both true and believed by some agent but this is not sufficient grounds for claiming that the agent knows the proposition since it may be believed for completely spurious reasons. Consequently, a theory of knowledge representation must be capable of representing and distinguishing between assertions of the form:

(1)   Agent A <u>believes</u> that p
(2)   Agent A <u>knows</u> that p
(3)   Agent A <u>believes</u> some <u>proposition</u> which is <u>false</u>

The upshot of these simple observations seems to be that any adequate theory of knowledge representation must in part be a *theory* of truth and modality. The central concern of such a theory must therefore be with the development of a language in which such assertions as (1), (2) and (3) can be represented together with the formulation of appropriate logics for such modal notions. Much of the present study is concerned with the development of such languages and logics. Not that we shall provide anything like a definitive answer. We shall not enter into debate about the logics which are most appropriate for any particular modality; this is already a well documented, even if controversial, area. Our goal is much less ambitious: we aim to draw certain boundaries around the possible form of such theories. The need for this stems from the particular nature of the theories we shall advocate. We firmly believe that the most natural and computationally tractable theory is a first-order one. This imposes limitations on the logics of modality and truth.

To some extent this goal might be taken as identical to that of natural language semantics, or rather that part of semantics concerned with *truth* and *modality*. One important aspect in which our task is different concerns the concentration in semantics on the systematic relation between syntactic structure and semantic representation. We are at liberty to concentrate on the language of semantic representation and only appeal to natural language as a guide since we are not primarily concerned with the syntax–semantics connection. Rather we are concerned with the language and nature of semantic representation. This is important since it enables us largely to divorce our study from considerations which relate directly to natural language syntax and how the syntactic or grammatical form of a sentence contributes to its logical representation. However, this separation is a delicate one and has to be exercised with care. In the end, our languages of knowledge representation must be as naturally expressive as natural language itself and the insights of natural language semantics cannot be ignored.

1.2 Propositions and Modality

One of the first questions which arises in such an endeavour concerns the arguments of these modal connectives: what is it that is believed, known or taken to be true? This is, of course, a non-trivial philosophical question and one that perhaps has no definitive answer. Nevertheless it is not one that can be avoided in any serious study of the issues at hand: the answer given to this question either explicitly or implicitly dictates the form of the theory that will be championed. Traditionally, *propositions* are taken to be the objects of belief, knowledge and truth. Of course, this is just a name to hang the problem on since now we are forced to face the question *what is a proposition*? Fortunately, the literature is full of possible answers and the formal theory of propositions largely determines the language and content of the resulting theory of modality. In what follows we shall explore various proposals from the AI and philosphical literature and see how they meet the needs of knowledge representation.

1.3 Possible Worlds

In one of the major paradigms, the notion of possible worlds plays the leading role. Propositions are taken to be sets of possible worlds and properties are understood as functions from individuals to propositions. The modal and doxastic operators are then treated as functions from propositions to propositions. In general, modal operators are analysed as functions which send a proposition P to that proposition which consists of all those worlds which are accessible from elements of P. Different choices of the relation of accessibility lead to different modal and doxastic notions (Hintikka [1962], Kripke [1963]).

This is the *classical* theory and forms the underlying semantic theory of first-order modal logic. The language of the latter is derived from the language of first-order logic by the addition of new sentential operators (the modal connectives) whose logic is given by one of the standard systems of modal logic. The different logics correspond to different properties of the relation of accessibility. In this approach, the modal connectives operate on sentences of the language where the latter are taken to denote propositions (i.e. sets of worlds).

The elegance of this simple theory together with the fact that the different modal and doxastic notions can be distinguished by varying the constraints on the accessibility relation are largely responsible for its abiding influence on both logicians and computer scientists. As it stands, however, it will not serve our purposes.

1.4 Higher-Order Modal Logic

This first-order approach will not be expressive enough for the goals of knowledge representation. For one thing there is a prima facie need to quantify over propositions and properties. Consider the sentences:

(4) Agent A believes a false proposition
(5) Every proposition agent A believes is false
(6) Agent A can perform every task agent B can

The expression of (4), (5) and (6) demands more than first-order modal logic since they involve quantification over propositions and properties and this is not available in first-order modal logic. Other reasons might be marshalled to justify such quantificational facilities. For example, in Ramsay[1988] the need for quantification over properties is illustrated by the desire to formulate *frame* axioms in a natural and succinct fashion. Here one has to stipulate those aspects of a situation which remain unchanged under a specified revision, and the natural and elegant way of achieving this involves quantification over properties.

   The introduction of such quantification moves us into the domain of higher-order intensional logic, the most highly developed version of which is due to Montague[1973]. In this logic we are able to quantify over propositions and properties and much else besides. The different kinds of objects are delineated by the formal notion of *type*: types are the basic ones [individuals (e) and truth values (t)], or function types (the type of functions from one type to a second), or the type of functions from the type of possible worlds (or some more complex index) to any existing type. Propositions are then analysed as functions from worlds to truth values (or equivalently sets of worlds) and properties as functions from individuals to propositions. The variables of the language are decorated with these types so, for example, the content of (4), (5), (6) can be formally expressed as

(4') $\exists x_{\langle w,t \rangle}(B_A(x) \,\& \sim (x))$
(5') $\forall x_{\langle w,t \rangle}(B_A(x) \rightarrow \sim (x))$
(6') $\forall x_{\langle i,\langle w,t \rangle \rangle}(\text{Perform}(B, x) \rightarrow \text{Perform}(A, x))$

where the variables $x_{\langle w,t \rangle}$ range over propositions and the $x_{\langle i,\langle w,t \rangle \rangle}$ over properties. In the higher-order intensional logics, quantification over individuals, propositions and properties is built into the language of the theory. Indeed, quantification over even higher type objects is permitted. This approach thus appears to be expressive enough for our needs but suffers from a possible conceptual drawback.

1.5   *Fine-Grained* Propositions

The major objection to the whole possible world approach, at least in regard to its application to the doxastic modalities such as belief and knowledge, concerns the nature of propositions within this regime. It is argued that propositions, as sets of possible worlds, are too *coarse-grained* to serve as arguments to the doxastic operators of knowledge and belief: if two sentences denote exactly the same set of possible worlds, then an agent who believes one is committed to believing the other. While this may be acceptable for certain notions of belief (e.g. rational belief) it does not seem persuasive for all notions. Mathematical belief seems to be a case in point. Since mathematical assertions are naturally taken to be necessarily true or necessarily false, under this account of belief an agent who believes one true assertion of mathematics (e.g. $2 + 2 = 4$) is thereby committed to believing all true assertions of mathematics. This seems not to supply an adequate account of mathematical belief. This criticism of the possible world approach to the attitudes of belief and knowledge naturally leads to the demand for a more *fine-grained* notion of proposition: one that will not commit an agent to believing all the logical consequences of his or her basic beliefs.

1.6   Propositions as Primitive

One proposal for overcoming this problem involves kicking away the possible world ladder which supports the notion of proposition. Propositions are then not unpacked in terms of possible worlds or any other supposedly more fundamental notion but taken as primitive. Subsequently, we are not forced to take the equality of propositions to be given by the extensional equivalence between sets of worlds. Rather, we are free to invest the notion of proposition with the properties we see fit. Thomason [1980] develops such an approach for higher-order intensional logic and thus combines the expressive power of Montague's intensional logic with a more *fine-grained* notion of proposition. In Thomason's simple type theory there are three basic types: e, t, and p, where e is the type of individuals, t the type of truth-values, and p the type of propositions. Higher-order types are constructed from these in the standard way. In addition, Thomason introduces a simple truth predicate to express the fact that a proposition is true. The Lambda Calculus (typed) is built into the system in the way familiar from Montague. Thomason's logic is quite complex and perhaps hard to work with in terms of the application at hand. Nevertheless, this is a step in the right direction.

1.7 Higher-Order Theories and Computational Tractability

Both the approaches of Montague and Thomason are versions of higher-order intensional logic. Indeed, it is exactly this quality which enables the expression of the logical content of sentences such as

(7) A believes everything B believes
(8) Something A believes is true

These can be expressed in the language of Montague's higher-order intensional logic (or with suitable modifications in Thomason's) by

(7') $\forall x_{\langle w,t \rangle}(\text{bel}(B, x) \rightarrow \text{bel}(A, x))$
(8') $\exists x_{\langle w,t \rangle}(\text{bel}(A, x) \ \& \ x)$

Unfortunately, this expressive power comes with a price tag. While it is true that first-order logic is only semi-decidable, higher-order logic is much worse. In first-order logic we can construct theorem provers which return a proof for any valid well-formed formulae even if they may fail to terminate when applied to invalid ones; in higher-order logic the theorem provers may fail to terminate even on the valid formulae. This is, of course, not a sufficient reason for discarding the higher-order approach and certainly not in the absence of a better alternative. But it is at least a reason for being somewhat circumspect about rushing headlong into higher-order logic.

1.8 Propositions as Sentences

A second and completely different proposal for the analysis of intensionality emanates largely from the AI community itself and seeks to view propositions as sentences in some language of semantic representation. Konolige [1986] and Moore [1980] implicitly seem to advocate such a view. This certainly addresses the fact that propositions need to be *fine-grained*, but there are obvious philosophical objections to such an approach. What is believed, known, deemed to be possible, etc. are not sentences in some language: sentences are just marks or symbols and the objects of knowledge or belief have semantic *content*; one cannot be said to believe a sentence but rather one stands in relation to its semantic content. Whatever the merits of this opinion, there are more devastating problems for this *syntactic* analysis of propositions.

6

Let L be the language in which such propositions are to be expressed. Then, writing $x \in L$ to indicate that x is a sentence of the language L, we can express:

(9) John believes something false

as

(9′) $\exists x(x \in L \ \& \ \text{bel}(\text{John}, x) \ \& \ \text{false}(x))$

This looks promising in that we have expressed the fact that John believes something false or believes a proposition which is false where propositions are identified with the sentence of L. But now we face a problem with regard to the language in which (9′) is expressed. This is not L and cannot be without violating predicativity. Suppose there are two believers, John and Peter, and Peter wishes to express the assertion that John believes something false. Then Peter's language must be expressive enough to express (9′) and consequently must have access to the *propositions* of John's language in that its variables of quantification must range over the sentences of L. One way of achieving this is to employ the Tarskian Object/Metalanguage distinction.

The main problem with such an approach concerns its expressive power. Consider the sentences:

(10) A believes that everything that B believes is true
(11) B believes that everything that A believes is true

Let O_A $(= M_{O_B})$ be the object language of A which also serves as the metalanguage for the object language O_B of B. Then we can attempt to express (10) as

(10′) $\text{bel}(A, \forall x \in O_B(\text{bel}(B, x) \rightarrow \text{true}(x)))$

where $x \in O_B$ has its obvious interpretation and facilitates quantification over the wff of B's language. In order to express (11), however, we require a language which has the facility to quantify over the the wff/beliefs of A. This cannot be achieved in O_B or O_A since we now wish to quantify over A's beliefs. We must resort to a metalanguage M_{O_A}. We can then attempt a statement of (11):

(11′) $\text{bel}(B, \forall x \in O_A(\text{bel}(A, x) \rightarrow \text{true}(x)))$

But now observe that (10′) does not include the belief of B expressed by (11′) since in O_A we can only quantify over those beliefs expressible in O_B. The belief expressed by (11′) is only expressible in a further metalanguage M_{O_A}. We can, of course, replace (10′) by (10″):

(10″) $\text{bel}(A, \forall x \in M_{O_A}(\text{bel}(B, x) \rightarrow \text{true}(x)))$

but then (11′) does not express what we think it does. Indeed, no matter how

7

far we climb up the object/metalanguage hierarchy we will not be able to capture the intuitive content of (10) and (11); some of the beliefs of A or B will always be left out. This whole approach runs into problems of expressive power: we are unable to express mutual belief and the reason stems from the hierarchical impositions placed on the organization of the languages of representation. This should be seen in contrast to the higher-order approach where the proposition variables range over the domain of propositions and include the denotations of the sentences expressed by (10) and (11). This is formally sanctioned by the comprehension schema of higher-order intensional logic which in particular guarantees that for every sentence of the formal language there is a proposition which furnishes its denotation.

This object/metalanguage analysis parallels the Tarski[1937] account of truth. In this theory there is a hierarchy of object/metalanguages and the metalanguage at each stage contains the truth predicate of the previous language. Tarski's theory of truth suffers similar drawbacks of expressive power. Moreover, there are obvious intuitive objections to this object/metalanguage approach. Natural language has no markings corresponding to these levels. According to this account every sentence must live somewhere in the hierarchy but given an arbitrary natural language sentence we seem unable to place it in such a hierarchy. Indeed, such cumbersome information is totally irrelevant to everyday communication. There are no good reasons to think that theories of knowledge representation, based upon such an explicit marking of levels, would be any less cumbersome and inefficient.

1.9 Syntactic Modality

One way out of this impasse is to remove the object/metalanguage distinction and identify the languages. The modal and truth operators now take sentences of the language (or their quoted relations) as arguments. Perlis[1988] has advocated such an approach to knowledge representation. This approach would certainly increase the expressive power of the language. Moreover, we would have a simple first-order system in which to express such modal notions since under this regime the modal operators are naturally analysed as simple first-order predicates. This appears to be in keeping with natural language itself. To see this consider the following sentences:

(12)   John believes that Peter sings
(13)   Mary thinks that Harry is a man
(14)   Peter knows that he will win
(15)   It is true that Peter believes that John sings

The most straightforward semantic account of such sentences would analyse the embedded clauses, under the government of the complementizer *that*, as singular terms. This offers the hope of a uniform treatment of all such *modal* sentences. Such an approach is defended at length in Bealer[1982] and Turner[1988]. We shall not here further pursue the semantic ramifications of such a view. We only note that this naive analysis seems worthy of consideration.

In order to see the impact of this proposal we need to spell it out in a little more detail. We assume a first-order language in which the truth and modal operators are predicates and the sentences of the language can occur as terms. Propositions are identified with the sentences of the language. In addition, if we wish to quantify over properties and other "higher-order" notions we require the language to be rich enough to support the definition of properties through Lambda abstraction: properties are analysed as propositional functions and given that sentences are to serve as propositions it is natural to unpack properties as the Lambda abstracts of such. All this can be simply achieved by including the Lambda Calculus as part of the syntax of terms. Indeed, once the Lambda Calculus is included in the language the wff can be coded as terms using the facilities of the Calculus.

Unfortunately, this approach is too expressive: it admits too liberal a notion of *proposition*. This can be easily demonstrated. The natural axioms of the truth predicate are given by the Tarski biconditionals:

(i) $A \leftrightarrow true(A)$ ...Tarski biconditionals

These presumably apply to all the propositions of the theory and, in so far as these are identified with the sentences of the language, to all the sentences. We now face a severe foundational problem. It arises because the facilities of the Lambda Calculus, which seem crucial to the application at hand, enable the derivation of a wff which says of itself that it is false:

(ii) $A \leftrightarrow {\sim} true(A)$ ...The *Liar*

This together with the Tarski biconditionals leads to inconsistency: in classical logic (i) and (ii) are formally inconsistent. Even if we had no truth predicate in the language we would not be safe since similar problems arise with the modal predicates if anything like the standard systems of modal logic are employed. The details of these derivations we provide later. For the present the result is all that is important.

1.10   Representationalism

This result indicates that the *syntactical* approach will not work and this is independent of any philosophical scruples we might have about the very idea of propositions being taken to *be* sentences. However, there is still one route we might explore. It is still open to us to maintain that although propositions are not sentences they are systematically related to them in that the structure of propositions exactly mirrors the structure of the sentences in some language of semantic representation. This is a weak *representationalist* view. Not only does this perspective address some of the philosophical objections raised against the simple-minded view of propositions *as* sentences but there is much to be said in favour of such a stance. The *fine-grained* notion of propositions we require certainly seems to demand that the identity criteria for propositions be as stringent as those for sentences. On its strongest interpretation this view commits us to a one-one correspondence between sentences and propositions. This is to be seen in contrast to the view that sentences are to be identified with propositions. It is philosophically important to distinguish between these two suggestions. To see that they are prima facie different, consider the analogy with set theory. Presumably, one would not be tempted to identify sets with the formal expressions in the language of set theory which we employ to represent them. The main reason appears to be that sets are extensional objects and the same set can be picked out by many different formal expressions. Nevertheless, this example of an axiomatic theory contains an important moral. It is not just the lack of a one-one correspondence between sets and the formal expressions of the language of set theory which blocks the identification. Formally it certainly does, but we would not be tempted to identify them even if the correspondence existed. Sets and the formal expressions of the language of set theory are conceptually very different and not to distinguish between them would be to adopt a formalist stance. The relationship is identical to that between numbers and numerals. Indeed, any axiomatic theory has to be expressed in a language and generally one is not tempted to identify the syntactic expressions of the theory with the "abstract" objects which are being referred to. So what is different about propositions and their linguistic counterparts? Apparently nothing except the existence of this one-one correspondence. Consequently, the temptation to identify them is stronger since the formal situation does not prevent it. However, it appears that there are positive reasons for treating propositions like sets or numbers, and like sets and numbers, propositions are best understood as "abstract" objects. Indeed, in any mathematical theory of propositions this is the only viable option.

Unfortunately, all this discussion appears to be somewhat beside the point. Thomason [1980] has shown conclusively that even this option is not open to us. Paradoxes still arise even under the guise of this more sophisticated representationalist view.

However, we have to proceed with care in drawing conclusions from these reflections on the representationalist approach to intensionality. What the paradoxes establish is not that we cannot have a *representationalist* view of propositions, but rather that *not all* sentences of our formal language can be representations of propositions. The language is too rich in its expressive power.

This point needs to be elaborated a little. The naive view is that there is a one-one correspondence between propositions and sentences: every sentence in our formal language determines precisely one proposition and vice-versa. This is the view which leads to inconsistency. The problem arises from the insistence that *every* sentence determines a proposition. We can still maintain a weak representationalist view of propositions where only a certain subclass of sentences determine propositions. This is indeed the approach we shall adopt. The theories we shall develop will be axiomatic theories of propositions and truth so that the notion of being a proposition will be characterized axiomatically rather than syntactically or grammatically.

1.11   Theories of Truth, Modality and Propositions

Even if all this is conceded, it still leaves us with the problem of developing suitable axiomatic theories of truth, propositions, properties and modality. The semantic and intensional paradoxes seem to block any simple and natural approaches. However, recent work on these paradoxes holds out some hope that some headway can be made. We briefly review some of this work as a preliminary to what follows. We shall concentrate on the theories of truth since, as we shall see, once this notion is in place the others follow suit.

In recent years there has been a revival in the development of semantic theories of truth. They are all attempts to develop theories of truth for languages which contain their own truth predicates, and, moreover, they are all semantic theories in that they all are grounded in some semantic interpretation of the truth predicate. We shall concentrate on those theories which are cast within the framework of classical logic. Three of the most influential theories are those of Scott[1975]–Aczel[1980]; Kripke[1975]–Gilmore[1974]–Feferman[1984] and Gupta[1982]–Herzberger[1982]. Here we briefly review them to set the scene for what follows.

Various approaches to the semantic paradoxes result in some logical schema to capture the safe instances of the Tarski biconditionals. For example, the approach of Gilmore[1974]–Feferman[1984] has as a consequence the schemata:

$$\text{true}(A) \quad \text{iff } A^+$$
$$\text{true}(\sim A) \quad \text{iff } A^-$$

11

where A^+ and A^- are in some sense *approximations* to A and $\sim$ A respectively. Even though the theory is cast within the general setting of classical logic the construction of the models is essentially based on a technique introduced into truth theory by Kripke[1975] and gains its formal credibility through a non-standard treatment of negation in wff such as A^+ and A^-. In essence these approximations are the result of pushing all negations into atomic position and replacing all such negations by *internal negations*. As a consequence the *internal* logic of the truth predicate is non-standard.

In contrast, the approach of Gupta[1982] and Herzberger[1982] is, in the semantic sense, totally classical. The Gupta–Herzberger theory of truth is based upon a notion of truth as revision. Herzberger calls his approach "naive semantics", referring to those naive beliefs about the concept of truth which lead to the paradoxes. The theory of truth advocated is a modification of Kripke's theory within a classical framework. The theory thus employs only classical models and ordinary two-valued valuations.

Aczel[1980] introduces the notion of a Frege Structure in an effort to capture the consistent subtheory of Frege's *Grundgesetze der Arithmetik*. Aczel formulates his theory within a model of the untyped Lambda Calculus and develops a theory of classes or types based upon a theory of truth and propositions. Frege Structures are models of the Lambda Calculus enriched with two subsets: a set of propositions and a set of true propositions. These sets satisfy very natural closure conditions with respect to the logical connectives: on the class of propositions the truth predicate obeys the Tarski criteria.

The first half of the book is given over to a detailed exposition of these three theories taken in the above order. We shall develop axiomatic theories of truth, propositions and properties which underlie these various semantic theories. In the second half of the book we turn to modal logic itself. After a brief excursion into traditional modal logic we develop theories of truth for first-order modal logic and revisit each of the theories within this new context. Finally, armed with these theories, we return to the development of logics of truth and *syntactic* or *predicative* modality.

2 Truth and Paradox

In this chapter we have two objectives. One is to set the technical scene for the book: we introduce the necessary background material on the Lambda Calculus and the Predicate Calculus. These provide the technical backbone of the theories we develop. The other objective concerns the formal derivation of the semantic paradoxes. We have already alluded to the fact that, in the context of the Lambda Calculus, the natural axioms for truth, given by the Tarski biconditionals, generate inconsistency. In this chapter we provide a formal account of this. This will lead us to the main theme of the first half of the book, namely the formulation and development of *logics of truth*. Of course, if the paradoxes did not arise, the Tarski biconditionals would supply the definitive logic of truth, but they do and so we have to be more circumspect in formulating such logics. Moreover, it is clearly not sufficient to simply formulate consistent logics; they have to be defended and developed in a philosophically principled fashion.

2.1 The Lambda Calculus

One foundational component of our formal theories is the Lambda Calculus. The machinery of the Calculus facilitates all the minor technical constructions we shall require. In addition, it supplies the means by which *properties* are defined through abstraction. The concept of *truth* enables the introduction of the concept of *proposition* and once this notion is in place the notions of *property* and *relation* will be introduced by means of the abstraction facilities of the Calculus: properties will be introduced as one-place propositional functions and, generally, n-place relations as n-place propositional functions.

We now present the technical details of the Calculus beginning with the syntax. The language of the Calculus is made up from constants, variables, abstractions and applications. More precisely, the language has the following content:

BASIC VOCABULARY
Individual variables $x, y, z, \ldots$

Individual constants   c, d, e, ...

INDUCTIVE DEFINITION OF TERMS
(i)   Every variable or constant is a term.
(ii)   If t is a term and x is a variable then $(\lambda x.t)$ is a term.
(iii)   If t and t' are terms then (tt') is a term.

Clause (ii) introduces the *abstractions* and (iii) the *applications*. Parentheses in (ii) and (iii) are used solely for indicating precedence. We shall adopt the standard conventions and, in particular, we write stt' for $((st)t')$ and $\lambda xy.t$ for $(\lambda x.(\lambda y.t))$, etc.

To facilitate the statement of the Lambda Calculus axioms we need the following technical definition of *substitution* and before this we need one other definition. In what follows $\approx$ will be employed for the syntactic identity of terms. The following definitions are quite standard but for completeness and future reference we include the details.

DEFINITION 2.0   (**freedom and bondage**)
Let t be any Lambda term and x a variable. Then x is *free* in t iff one of the following holds:

(i) $x \approx t$
(ii) $t \approx (ss')$ and x is free in s or x is free in s'
(iii) $t \approx \lambda y \cdot s$ and $x \neq y$ and x free in s.

In contrast we define x is **bound** in t by

(i) $t \approx \lambda y \cdot s$ and $x \approx y$ or x is bound in s
(ii) $t \approx (ss')$ and x is bound in s or x is bound in s'.

With these notions in place we can present the main definition.

DEFINITION 2.1   (**substitution**)
Let s and t be terms and x a variable. Then define $t[s/x]$, *the result of substituting s for every free occurrence of x in t* by recursion, as follows:

(i)   If $t \approx x$ then $t[s/x] \approx s$
       If $t \approx b$ (a variable or constant $\neq x$) then $t[s/x] \approx t$
(ii)   If $t \approx (t't'')$ then $t[s/x] \approx (t'[s/x]t''[s/x])$
(iii)   If $t \approx \lambda y.t'$ then there are two cases:
       (a)   if $x \approx y$ then $t[s/x] \approx t$
       (b)   if $x \neq y$ then there are two subcases:
       (b1)   if y is not free in s or x is not free in t' then

$$t[s/x] \approx \lambda y.t'[s/x]$$

(b2) if y is free in s and x is free in t' then

$$t[s/x] \approx \lambda z.t'[z/y][s/x]$$

where z is not free in s.

The only non-obvious case is given by clause (iii)b which splits into two cases. To justify its presence consider the instance of substitution:

$$(\lambda y.x)[w/x]$$

Intuitively, this ought to be the constant function with value w (i.e. $\lambda y.w$), but without the split between b1 and b2 it will denote the identity function ($\lambda w.w$) when $w \approx y$; hence the complication.

With these technical definitions out of the way we can proceed to the axioms and rules themselves. We adopt the following standard axiomatization of the $\lambda\beta$-Lambda Calculus:

AXIOM-SCHEMATA OF THE $\lambda\beta$-CALCULUS

(α) $\lambda x.t = \lambda y.t[y/x]$ y not free in t
(β) $(\lambda x.t)t' = t[t'/x]$
(ρ) $t = t$

RULES OF $\lambda\beta$-CALCULUS

(μ) $\dfrac{t = t'}{st = st'}$

(ν) $\dfrac{t = t'}{ts = t's}$

(ξ) $\dfrac{t = t'}{\lambda x.t = \lambda x.t'}$

(τ) $\dfrac{t = t' \qquad t' = s}{t = s}$

(σ) $\dfrac{t = s}{s = t}$

We shall write $\lambda\beta \vdash t = s$ iff $t = s$ is provable from the axiom schemata and rules of the $\lambda\beta$-Calculus. Let $\lambda\beta = \{s = t : \lambda\beta \vdash s = t\}$. In other words, $\lambda\beta$ is the set of all the equality assertions derivable in the theory $\lambda\beta$. This double use of $\lambda\beta$ should not cause confusion.

15

The first axiom (α) just allows for the renaming of bound variables, while the second (β) captures the obvious intuitions governing functional application. The rules close off the axioms under abstraction and application.

In addition, we require a few basic facts about the Lambda Calculus. We shall be brief and refer the reader to Hindley and Seldin [1986] and Barendregt [1984] for more details. We shall assume some standard representation of the pairing and projection combinators $\langle . , . \rangle$, fst, snd which satisfy: $\mathrm{fst}(\langle x, y \rangle) = x$ and $\mathrm{snd}(\langle x, y \rangle) = y$; n-tuples for $n > 2$ can then be represented by iterating this construction. We shall employ some standard representation of the numerals (e.g. the Church representation) $0, 1, 2, 3, 4, 5, \ldots$ etc. Finally, we shall also appeal to the fixpoint theorem of the Lambda Calculus which takes the following form.

THEOREM 2.2
There is a Lambda term **Y** such that for every Lambda term t,

$$\lambda\beta \vdash \mathbf{Y}t = t(\mathbf{Y}t)$$

PROOF Put $\mathbf{Y} = \lambda f \cdot (\lambda x \cdot f(xx))(\lambda x \cdot f(xx))$, then observe that

$$\begin{aligned}
\mathbf{Y}t &= (\lambda x \cdot t(xx))(\lambda x \cdot t(xx)) \\
&= t((\lambda x \cdot t(xx))(\lambda x \cdot t(xx))) \\
&= t(\mathbf{Y}t) \quad \bullet
\end{aligned}$$

This completes our brief excursion into the untyped Lambda Calculus. Fortunately, the above facts are all that we shall require. We hope that the brevity of our exposition will not prove too much for the uninitiated reader. We shall not employ anything outside this fragment but more details can be found in Hindley and Seldin [1986].

2.2 Models of the Lambda Calculus

All the model-theoretic constructions we shall pursue in later chapters will take place within the setting of Lambda Calculus models: all the basic model structures of all the languages we shall employ will be built from models of the Lambda Calculus. In this section we outline the original approach to such models due to Dana Scott [1973]. There are more abstract notions of model which are less mathematically complex but we feel that the price of a little mathematical sophistication is well compensated for by the concreteness of the following structures. It should be stressed, however, that nothing of what follows depends upon our particular choice of model structure. We present the following only for concreteness and because self-reference is one of the components which leads to paradox. Scott models provide a fine mathematical view of this notion.

Our models of the Calculus are built from the following class of mathematical structures.

DEFINITION 2.3
A *domain* is a partially ordered set, with a least element u, and where every ω-sequence has a least upper bound.

We spell this out in a little more detail. Let D be a domain and let $\subseteq$ be the ordering of the domain. The element u is the *least element* of the domain D if $u \subseteq d$ for each d in D. An ω-sequence is any sequence of the form $d_0 \subseteq d_1 \subseteq \ldots \subseteq d_n \subseteq \ldots$ where $d_i \in D$ for $i \geqslant 0$. An element d is an *upper bound* of the sequence if $d_i \subseteq d$ for each $i \geqslant 0$; it is a *least upper bound* if $d \subseteq d'$ for any other upper bound d' of the sequence. We write the least upper bound of the sequence as $\underset{i}{\cup} d_i$.

The next notion concerns the *acceptable* functions between such domains. The functions which provide the denotations of Lambda terms in the model are of a restricted kind.

DEFINITION 2.4
A function $f: D \to D'$ is *continuous* iff for each ω-sequence $\langle d_n \rangle_{n \in \omega}$ in D,

$$f\left(\underset{i}{\cup} d_i\right) = \underset{i}{\cup} f(d_i)$$

The next definition indicates how the class of continuous functions from D to D' (where D and D' are domains) forms a domain.

DEFINITION 2.5
Let $[D \to D']$ be the class of continuous functions from D to D'. For f, $g \in [D \to D']$ define:

$$f \subseteq g \leftrightarrow (\forall d \in D)(f(d) \subseteq' g(d))$$

where $\subseteq'$ is the ordering of D'.

The next theorem informs us that this construction preserves the notion of being a domain.

THEOREM 2.6
$[D \to D']$ with the above ordering forms a domain.

PROOF   The least element is the function f given by $f(d) = u'$ for each d in D, where u' is the least element of D'. Least upper bounds are defined by

$$\left(\bigcup_i f_i \right)(d) = \bigcup_i (f_i(d))$$

where the second least upper bound is computed in D'. We leave the reader to check the continuity properties.   ●

This completes the basic preliminaries. In the present context a *Scott model* of the Lambda Calculus is a domain D which is isomorphic to its own continuous function space, where the isomorphisms are themselves continuous. More precisely:

DEFINITION 2.7
A *Scott model* is a triple $\mathcal{D} = \langle D, \Phi, \Psi \rangle$ where

   (i) D is a domain
   (ii) $\Phi : D \to [D \to D]$ and $\Psi : [D \to D] \to D$ are continuous isomorphisms.

We next provide a brief sketch of the construction of a Scott model. The reader who is prepared to take the existence of such a model on trust should move immediately to section 2.3 and the semantic clauses for the Lambda Calculus. We shall not give all the details of the construction but enough for the interested reader to complete the proof.

The basic idea behind the Scott construction is to build a sequence of domains, induced by applying the function space construction, and then (in some sense) proceed to the limit. We then obtain a domain D isomorphic to its own continuous function space.

The first step in the construction of the domain D is to define the sequence of function space domains. Let E be any domain. We define a sequence of domains based upon E by induction:

$$E_0 \quad = E$$
$$E_{n+1} = [E_n \to E_n], \quad n \geqslant 0$$

In order to proceed to the "limit" we must indicate how to embed each E_n in E_{n+1}. To this end we introduce a sequence of continuous functions:

$$p_n : E_n \quad \to E_{n+1}$$
$$q_n : E_{n+1} \to E_n$$

by induction on $n \geqslant 0$. For the base step we define p_0 and q_0 as follows:

$$p_0(a) \quad = \lambda e \in E.a \qquad \text{for } a \text{ in } E$$

18

$$q_0(x') = x'(u) \qquad \text{for } x' \in [E \to E]$$

Inductively, assume that p_n and q_n have been defined. Then define

$$p_{n+1}(x) = p_n \circ x \circ q_n \qquad \text{for } x \text{ in } [E_n \to E_n]$$
$$q_{n+1}(x') = q_n \circ x' \circ p_n \qquad \text{for } x' \text{ in } [E_{n+1} \to E_{n+1}]$$

where $\circ$ represents functional composition. These functions (as ·an easy inductive argument shows) are continuous and satisfy $q_n(p_n(f)) = f$ and $p_n(q_n(g)) \subseteq g$. Technically they form a "projection" pair. In particular, the function p_n is one-to-one. This permits us to view E_n as a subdomain of E_{n+1} (under the projection p_n). In fact, we can extend the mappings p_n, q_n to mappings (continuous) $p_{nm} : E_n \to E_m$ as follows:

$$p_{nm}(f) = \begin{cases} p_{m-1} \circ \ldots \circ p_n & n < m \\ f & n = m \\ q_m \circ \quad \ldots \circ q_{n-1} & m < n \end{cases}$$

Once again it is easy to check that $p_{mn}(p_{nm}(f)) = f$ and $p_{nm}(p_{mn}(g)) \subseteq g$ for $0 \leqslant n \leqslant m$.

All this facilitates the construction of the domain of a Scott model. As we indicated before we employ the sequence of domains to construct a domain isomorphic to its own continuous function space. The following is the notion of "limit" alluded to.

DEFINITION 2.8
The domain D consists of the set

$$\{\langle f_n \rangle_{n \geqslant 0} : f_n \in E_n \ \& \ q_n(f_{n+1}) = f_n\}$$

where for $\langle f_n \rangle$, $\langle g_n \rangle$ in D, $\langle f_n \rangle \subseteq \langle g_n \rangle \Leftrightarrow (\forall n)(f_n \subseteq g_n)$.

Under this ordering D forms a domain. Moreover, we can embed each of the E_n ($n \geqslant 0$) in D by defining $\Phi_n : E_n \to D$ and $\Psi_n : D \to E_n$ as follows:

$$\Phi_n(f) = \langle p_{nk}(f) \rangle_{k \geqslant 0}$$
$$\Psi_n(f) = f_n$$

Once more we leave the reader to check that $\Psi_n(\Phi_n(f)) = f$ and $\Psi_n(\Phi_n(f')) \subseteq f'$. We can, therefore, regard each E_n as a subdomain of D under the embedding $\Phi_n : E_n \to E$. This permits us to formally identify each $d \in E_n$ with $\Phi_n(x)$ and to identify each element of E_n with an element of D.

The following lemmas are standard; the proofs can be found in Barendregt [1984].

LEMMA 2.9

For each $f \in D$

    (i)   If $f \in E_n$ then $f = f_n$

    (ii)   If $f \in E_n$ then $p_n(f) = f$

    (iii) If $f \in E_{n+1}$ then $q_n(f) \subseteq f$

LEMMA 2.10

For each $f \in D$

    (i) $(f_n)_m = f_{\min(n,m)}$

    (ii)   If $m \geqslant n$ then $f_n \subseteq f_m \subseteq f$

    (iii) $f = \bigcup_n f_n.$

The domain component of our Scott model is now in place. We turn next to the construction of the functions which move us between D and its continuous function space.

LEMMA 2.11

Define application in D by $\mathrm{app}(f, e) = \bigcup_n f_{n+1}(e_n)$. Then application in D is

continuous and satisfies

$$\mathrm{app}(f_{n+1}, e) = \mathrm{app}(f_{n+1}, e_n) = [\mathrm{app}(f, e_n)]_n$$

For the proof we once again refer the reader to Barendregt [1984]. This lemma and the next provide the functions of a Scott model.

LEMMA 2.12

For each $f \in [D \to D]$ there exists an x_f in D such that for each e in D,

$$f(e) = \mathrm{app}(x_f, e)$$

The x_f in question is given by $x_f = \bigcup_n \lambda d \in D.(\mathrm{app}(f, d)_n)$. This follows by a

relatively straightforward computation.

THEOREM 2.13

The structure $\langle D, \Phi, \Psi \rangle$ is a Scott model where $\Phi : D \to [D \to D]$ and $\Psi : [D \to D] \to D$ are given by

$$\Phi(d) = \lambda e \in D.\mathrm{app}(d, e) \quad \text{and} \quad \Psi(f) = x_f$$

From now on we shall work with a fixed Scott model $\mathscr{D}$. As we have said earlier, nothing we do depends upon the actual nature of the models: we are only interested in their abstract form which is given by 2.7. The reader new to this subject may have found this section heavy going. We have only attempted to give a sketch. More details can be found in the literature cited.

2.3   Semantics of the Lambda Calculus

Given such a structure we can interpret the Lambda Calculus in it in a reasonably straightforward way. The semantics will be given relative to an assignment function g which assigns elements of D to variables and an interpretation function i which assigns elements of D to constants. We shall employ the notation $g(d/x)$ for that assignment function identical to g except that d is bound to x. For convenience we drop all reference to $\mathcal{D}$ in the semantic definition which follows. Also from here on we shall assume that i is fixed.

L1 $\mathscr{I}[x]_g = g(x)$

L2 $\mathscr{I}[c]_g = i(c)$

L3 $\mathscr{I}[\lambda x.t]_g = \Psi(\lambda d.\mathscr{I}[t]_{g(d/x)})$

L4 $\mathscr{I}[t(t')]_g = \Phi(\mathscr{I}[t]_g)(\mathscr{I}[t']_g)$

The lambda abstraction employed in L3 is, of course, part of the metalanguage. We might unpack L3 differently as

L3 $\mathscr{I}[\lambda x.t]_g = \Psi(f)$
     where for each $d \in D$, $f(d) = [t]_{g(d/x)}$

This avoids the use of the lambda in the metalanguage by the employment of the more familiar mathematical vernacular.

These clauses are all standard. In L3 the function Ψ is used to pull the function back from $[D \rightarrow D]$ to D and in L4 the function Φ is used to create a function for the application. Thus all elements of the language obtain their denotations in D. The important point is that the functions $\lambda d.\mathscr{I}[t]_{g(d/x)}$ are all members of the continuous function space $[D \rightarrow D]$ and so the definition is sound. This requires proof but it follows by an easy inductive argument. In passing it is worth saying that one can provide a more abstract notion of model as a triple $\langle D, \Phi, \Psi \rangle$ where Φ and Ψ are isomorphisms and the space $[D \rightarrow D]$ is closed under Lambda abstraction, but this depends too much on the syntax of the Lambda Calculus. We refer the reader to Hindley and Seldin [1986] and Barendregt [1984] for more details of Lambda Calculus Models.

We shall say that an equation, $t = s$, between terms is *valid* in such a Scott model iff for all assignment functions g, $[t]_g = [s]_g$.

THEOREM 2.14

If $\lambda\beta \vdash t = s$ then $t = s$ is valid in all Scott models.

This can be established by induction on the axioms and rules of the $\lambda\beta$-Calculus. Once again we refer the reader to the literature for details.

2.4 The Predicate Calculus

One plank in the formal framework is now in place. The second concerns the language of well-formed formulae (wff). This language (L) has three types of atomic wff: $t = s$, $T(t)$ and $F(t)$. The first is equality of terms, the second asserts that a term is true, and the third asserts that a term is false. The terms are those of the Lambda Calculus.

INDUCTIVE DEFINITION OF WFF

 (i) If t is a term then $T(t)$ and $F(t)$ are wff.

 (ii) If t and t' are terms then $t = t'$ is a wff.

 (iii) If A and B are wff then so are $A \& B$, $A \vee B$, $\sim A$, $A \rightarrow B$.

 (iv) If x is a variable and A is a wff then $\forall xA$ and $\exists xA$ are wff.

We shall employ $\perp$ as an abbreviation for some absurd statement and $A \leftrightarrow B$ as an abbreviation for $A \rightarrow B \& B \rightarrow A$. We adopt some standard axiomatization of first-order logic with equality. We shall write $LC \vdash A$ if A is provable in classical first-order logic with equality from the theory $\lambda\beta$ (i.e. the theory which consists of all provable assertions in the theory of the $\lambda\beta$-Calculus).

This brings us to the notion of a model for the language L. It is standard apart from the fact that the extensions of truth and falsity are singled out for special treatment.

DEFINITION 2.15

A *model* for L is a tuple $\mathcal{M} = \langle \mathcal{D}, T, F \rangle$ where $\mathcal{D}$ is a model of the Lambda Calculus and

$$T: D \rightarrow \{0, 1\} \quad \text{and} \quad F: D \rightarrow \{0, 1\}$$

The wff of the language L can now be given truth conditions in the traditional way where T and F respectively provide the extensions of the truth and falsity predicates.

$$\mathcal{M} \vDash_g s = t \quad \text{iff} \quad \mathcal{I}[t]_g = \mathcal{I}[s]_g$$
$$\mathcal{M} \vDash_g T(t) \quad \text{iff} \quad T(\mathcal{I}[t]_g) = 1$$

$\mathcal{M} \vDash_g F(t)$ iff $F(\mathcal{I}[t]_g) = 1$

$\mathcal{M} \vDash_g A \& B$ iff $\mathcal{M} \vDash_g A$ and $\mathcal{M} \vDash_g B$

$\mathcal{M} \vDash_g A \vee B$ iff $\mathcal{M} \vDash_g A$ or $\mathcal{M} \vDash_g B$

$\mathcal{M} \vDash_g A \rightarrow B$ iff $\mathcal{M} \vDash_g A$ implies $\mathcal{M} \vDash_g B$

$\mathcal{M} \vDash_g \sim A$ iff not $\mathcal{M} \vDash_g A$

$\mathcal{M} \vDash_g \forall x A$ iff for all d in D, $\mathcal{M} \vDash_{g(d/x)} A$

$\mathcal{M} \vDash_g \exists x A$ iff for some d in D, $\mathcal{M} \vDash_{g(d/x)} A$

The concept of validity is also quite standard.

DEFINITION 2.16
A wff A of L is *valid in a model* $\mathcal{M}$ iff $\mathcal{M} \vDash_g A$ for all assignment functions g.

The presentation so far is perfectly orthodox. We have done nothing except give fairly standard expositions of the Lambda Calculus and the Predicate Calculus. In particular, we have imposed no conditions on the extensions of truth and falsity. As a consequence, the theory is completely useless for reasoning about truth since there are no axioms for the truth predicate. Indeed, before we can even state any axioms, we must first indicate how the wff can be treated as terms so that the truth predicate T can be applied to them. We therefore add a new clause to the language:

(v)   If A is a wff then ˆA is a term.

Actually, we do not have to add this as a new clause since we can achieve the same effect by coding. Using the pairing/triple combinator and numerals of the Lambda Calculus we can code the wff as terms of the Lambda Calculus as follows:

$$
\begin{aligned}
&\hat{}(x) & &= x \\
&\hat{}(c) & &= c \\
&\hat{}(ts) & &= \hat{}(t)\hat{}(s) \\
&\hat{}(\lambda x.t) & &= \lambda x.\hat{}(t) \\
&\hat{}(t = s) & &= \langle 0, \hat{}(t), \hat{}(s) \rangle \\
&\hat{}(T(t)) & &= \langle 1, \hat{}t \rangle \\
&\hat{}(F(t)) & &= \langle 2, \hat{}t \rangle \\
&\hat{}(\sim A) & &= \langle 3, \hat{}A \rangle \\
&\hat{}(A \& B) & &= \langle 4, \hat{}A, \hat{}B \rangle \\
&\hat{}(A \vee B) & &= \langle 5, \hat{}A, \hat{}B \rangle \\
&\hat{}(A \rightarrow B) & &= \langle 6, \hat{}A, \hat{}B \rangle \\
&\hat{}(\forall x A) & &= \langle 7, x, \hat{}A \rangle \\
&\hat{}(\exists x A) & &= \langle 8, x, \hat{}A \rangle
\end{aligned}
$$

The details of this coding are not important; indeed, the only significant point about ˆ is that ˆ(A) and ˆ(B) will only be the same when A and B are identical wff, i.e. the representation enjoys a certain *independence property*. Moreover this property is inherited by the models, i.e. if $\mathcal{I}[\hat{\ }(A)]_g = \mathcal{I}[\hat{\ }(B)]_g$ then A and B will be the same wff. In the model we shall let CODE = $\{\mathcal{I}[\hat{\ }A]_g$: A is a wff and g an assignment function$\}$. In future we shall assume some such *independent* coding for our language and its extensions without further discussion of the details.

2.5   The Tarski Biconditionals

We now come to our promised account of the formal derivation of the semantic paradoxes. The intuitive principle which governs the logic of the truth predicate is given by the Tarski biconditionals:

TB $T(A) \leftrightarrow A$

where T(A) is an abbreviation for T(ˆA). Unfortunately, this theory is inconsistent. This stems directly from the fixpoint property of the Lambda Calculus. We shall refer to the following as the *diagonalization* lemma.

LEMMA 2.16
Let $A(x)$ be any wff whose only free variable is x. Then there is a sentence B such that

$$\text{LC} \vdash B \leftrightarrow A[\hat{\ }B/x]$$

   PROOF   Let $f = \lambda x.\hat{\ }A(x)$ and $t = \mathbf{Y}f$. Then by the fixpoint theorem we have $t = \hat{\ }A[t/x]$. By the equality rule of the Predicate Calculus we have

$$A[t/x] \leftrightarrow A[\hat{\ }A[t/x]/x]$$

Finally put $B = A[t/x]$.   ●

A simple application of this result gives us the infamous paradox of the liar: the diagonalization lemma enables the construction of a sentence which asserts its own falsity.

COROLLARY 2.17   LC + TB is inconsistent.

   PROOF   Let $A(x) = \sim T(x)$. Then by the theorem there exists a wff B such that $B \leftrightarrow A[\hat{\ }B/x]$, i.e. $B \leftrightarrow \sim T(B)$. By the Tarski biconditionals $\sim T(B)$ is equivalent to $\sim B$, hence we have a contradiction.   ●

As a matter of interest notice that other paradoxes are derivable in the setting of the Lambda Calculus. The facility for abstraction available in the Lambda Calculus enables the derivation of the *Russell* paradox without explicit appeal to the fixpoint combinator.

Russell   Write $\{x:B\}$ for $\lambda x.B$ and $x\varepsilon y$ for $T(xy)$. Let $t = \{x: \sim(x\varepsilon x)\}$ and then put $A = (t\varepsilon t)$. Now tt is equal by β-reduction to $\sim(t\varepsilon t)$. By the equality rule we have $T(tt) \leftrightarrow T(\sim(t\varepsilon t))$, i.e. $A \leftrightarrow T(\sim A)$ which by the Tarski biconditionals yields the equivalence: $A \leftrightarrow \sim A$.

Thus the Tarski biconditionals generate no end of problems. The urgent question is: which logics of truth are available to us? To address this question we need to examine *more carefully* the Tarski biconditionals.

2.6   Logics of Truth

The *logic of truth* defined by the Tarski biconditionals can be equivalently formulated in terms of the classical truth conditions:

(1) $T(A)$ $\leftrightarrow A$ for atomic A
(2) $T(A \& B)$ $\leftrightarrow T(A) \& T(B)$
(3) $T(\sim A)$ $\leftrightarrow \sim T(A)$
(4) $T(\forall x A)$ $\leftrightarrow \forall x T(A)$

This defines the *logic of truth* which we refer to as **CL** (and which we know to be inconsistent). In the next three chapters we shall be concerned with the development of *logics of truth* which are consistent. We first make some general observations regarding the scope and limitations of our task. We impose two constraints on our logics of truth which will remain intact for all the theories we shall study:

DIS $\sim(T(t) \& F(t))$
FT $F(A) \leftrightarrow T(\sim A)$

The principle DIS denies that any object can be in both the extensions of truth and falsity, while FT is definitional in content and insists that a sentence is false precisely when its negation is true. These are entirely unproblematic and, indeed, necessary for any sensible theory of truth.

One of the principles 1–4 which constitutes **LC** must be given up. In all the theories which we study axiom (3) is discarded. Put slightly differently, we shall abandon the principle of *bivalence*:

PB $T(A) \vee F(A)$

Not everything will be true or false or, more evocatively, not everything will be a *proposition*, where we have surreptitiously introduced the notion of *proposition*.

DEFINITION 2.19

We define the predicates **proposition** and **paradoxical** by

$$P(x) =_{\text{def}} T(x) \vee F(x) \quad \text{and} \quad U(x) =_{\text{def}} {\sim} P(x)$$

Propositions are those objects which are true or false and paradoxical objects are those which are neither. Of course, abstract definitions like this are no explanation. We have to see how these predicates get semantically and formally fleshed-out in the theories which follow. It is in this sense that a theory of *truth* supports a theory of *propositions*, which in its turn supports the notions of *property* and *relation*.

As an aside it is worth noting that there is a curious asymmetry between the two components of the Tarski biconditionals:

$$T(A) \rightarrow A \qquad \text{(i)}$$
$$A \rightarrow T(A) \qquad \text{(ii)}$$

If we admit *all* instances of (ii) then, in classical logic at least, all instances of (i) follow. The proof is set out below.

(1)	$T(A)$	...	assumption
(2)	${\sim}A$	...	assumption
(3)	$T({\sim}A)$	...	from (2) and principle (ii)
(4)	$T(A) \& T({\sim}A)$	...	from (3) and (1)
(5)	Contradiction	...	from (4), DIS and FT
(6)	A	...	from (2), discharging (2)
(7)	$T(A) \rightarrow A$	...	from (1), discharging (1).

This is indeed curious since as a consequence we have that DIS + FT + (ii) implies (i) and hence all the Tarski biconditionals follow. Consequently, even this apparently rather weak theory is inconsistent. The alternative theory DIS + FT + (i) does not imply (ii) and is, as we all see later, consistent.

This completes our discussion of the background material necessary for the exposition of the logics which follows. Our intention is to review the various semantic theories of truth that have been developed and explore their underlying *logics of truth*.

3　Truth through Fixpoints

The *logic of truth* enshrined in the Tarski biconditionals is classical and this is one of the components which leads to inconsistency. Our way out involves tinkering with the logic of the truth predicate and in this chapter we examine the route which leads us through the terrain of three-valued logic. In fact, many suggestions for a resolution of the semantic paradoxes amount to little more than the claim that certain sentences are neither true nor false. Bochvar [1939], for example, advocated a three-valued logic in which the third value, "paradoxical", is attached to the recalcitrant sentences. The paradoxes are then blocked.

As it stands, such a proposal is not entirely satisfactory. I take it that any satisfactory account of the paradoxes has to fulfil two requirements. Firstly, it has to be formally adequate in that it must block the formal derivation of paradox and, moreover, this has to be done with a degree of caution since one does not want to cripple ordinary reasoning. Secondly, it must be philosophically adequate: any principle that is rejected must be accompanied by independent reasons for the rejection. It is not sufficient to put aside a principle simply because doing so renders the system consistent. Ideally, there ought to be some independent conceptual explanation of why the principle is to be abandoned. The naive rejection of classical logic in favour of three-valued logic arguably fails the first test since three-valued logic appears not to reflect ordinary reasoning. More importantly, even if it provides a formal solution, without some semantic explanation of why three-valued logic holds the key to resolution we have little more than a technical device for avoiding paradox.

Fortunately, the three-valued approach has been given new life by recent model-theoretic accounts of truth. Kripke [1975], building upon Kleene strong three-valued logic, has developed a sophisticated proposal which offers a model-theoretic account of the notion of "paradoxical". A similar attack on the problem has been made by Woodruff [1975] and Feferman [1984] and, even earlier, in connection with the set-theoretic paradoxes, by Gilmore [1974]. Perlis [1988] advocates the application of Gilmore's theory to knowledge representation.

In this chapter we shall develop a *logic of truth* based upon Kleene strong three-valued logic. This does not mean that we are giving up classical logic in favour of a three-valued logic but rather the *logic of truth* itself is to be based upon Kleene logic. As a consequence, we abandon the principle of bivalence:

PB $T(A) \vee F(A)$

but not the law of excluded middle:

LEM $A \vee \sim A$

Subsequently, not every assertion will formally be rendered true or false under the government of the truth predicate, or equivalently, not every assertion will be a *proposition*. Consequently, we are not forced to discard classical logic. It should also be stressed that, although opting for a three-valued logic of truth means that PB must be put aside, the converse is false. There are other moves which can be made which abandon PB but which do not lead to three-valued theories in the traditional sense. In fact, in the next chapter we shall discuss various *modal logics of truth* which do not support PB. With regard to the present theory, we shall first state the theory and then develop those semantic intuitions which furnish its conceptual foundation.

3.1 Kleene Strong Three-Valued Logic as a Theory of Truth and Falsity

The Kleene strong three-valued truth conditions for negation, conjunction and disjunction are given by the following truth tables where t, f, u represent *truth*, *falsity* and *undefined* respectively.

Negation			Conjunction				Disjunction			
$\sim$			$\&$	t	f	u		$\vee$	t	f	u
t	f		t	t	f	u		t	t	t	t
f	t		f	f	f	f		f	t	f	u
u	u		u	u	f	u		u	t	u	u

Intuitively, the value u is not to be taken as a third truth value but rather to indicate that the wff is neither true nor false. These truth tables thus reflect the semantic perspective that certain wff may be neither true nor false. Nevertheless, they are in complete harmony with the classical tables in that, where all the truth values of the components of a wff are defined, the result will be. Indeed, they go further in that they strive to return the classical value wherever possible. For example, the disjunction of two wff is rendered true by the truth of either, regardless of whether the other disjunct is undefined. As a consequence the truth tables can be stated in a rather elegant way which makes no explicit mention of the undefined value:

~ A	is True	iff A is False
~ A	is False	iff A is True
A & B	is True	iff A is True and B is True
A & B	is False	iff A is False or B is False
A ∨ B	is True	iff A is True or B is True
A ∨ B	is False	iff A is False and B is False

The first version of the theory we shall study in this chapter is nothing more than a formal statement of these truth conditions. The following axioms exactly capture the Kleene truth conditions for the Predicate Calculus.

AXIOMS OF **KFG**

F1	$T(A \& B) \leftrightarrow T(A) \& T(B)$
F2	$F(A \& B) \leftrightarrow F(A) \vee F(B)$
F3	$T(A \vee B) \leftrightarrow T(A) \vee T(B)$
F4	$F(A \vee B) \leftrightarrow F(A) \& F(B)$
F5	$T(\sim A) \leftrightarrow F(A)$
F6	$F(\sim A) \leftrightarrow T(A)$
F7	$T(A \rightarrow B) \leftrightarrow F(A) \vee T(B)$
F8	$F(A \rightarrow B) \leftrightarrow T(A) \& F(B)$
F9	$T(\forall xA) \leftrightarrow \forall xT(A)$
F10	$F(\forall xA) \leftrightarrow \exists xF(A)$
F11	$T(\exists xA) \leftrightarrow \exists xT(A)$
F12	$F(\exists xA) \leftrightarrow \forall xF(A)$

In addition to these axioms we need to reflect the idea that truth and falsity are disjoint, i.e. no sentence can be both true and false. This is the content of the following:

DIS $\sim (T(A) \& F(A))$

This is built into all our theories, as indeed is F5. Finally, we provide the obvious truth conditions for the atomic assertions of L which in particular concern the behaviour of atomic sentences under T and F and sanction the derivation of the Tarski biconditionals for such assertions.

A1	$T(x) \leftrightarrow T(T(x))$
A2	$F(x) \leftrightarrow F(T(x))$
A3	$F(x) \leftrightarrow T(F(x))$

A4 $T(x) \leftrightarrow F(F(x))$
A5 $T(x = y) \leftrightarrow x = y$
A6 $F(x = y) \leftrightarrow x \neq y$

KFG is the logic F1 − F12 + DIS + A1 − A6; it exactly reflects the Kleene truth conditions for L.

Since LEM (the law of excluded middle, $A \lor \sim A$) fails for Kleene logic, PB fails for the logic of truth, **KFG**. Recall that we introduced the notion of *proposition* formally by the definition $P(x) =_{\text{def}} T(x) \lor F(x)$. In particular, a wff A will be a proposition precisely when $T(A)$ or $F(A)$. The failure of bivalence means that not every wff will be a proposition and, in particular, the *liar* will not be.

It is worth observing that other three-valued logics such as Bochvar's can be employed to induce logics of truth. Since Bochvar originally advocated his logic to deal with the paradoxes it seems fitting to illustrate matters by employing it. Bochvar's logic is internalized to yield a logic of truth as follows. First define

$$U(x) =_{\text{def}} \sim P(x)$$

to mean x is *paradoxical*. Bochvar's logic can then be formalized as follows:

$T(A \& B) \leftrightarrow T(A) \& T(B)$
$U(A \& B) \leftrightarrow U(A) \lor U(B)$
$F(A \& B) \leftrightarrow (P(A) \& F(B)) \lor (P(B) \& F(A))$
$T(A \lor B) \leftrightarrow (T(A) \& P(B)) \lor (T(B) \& P(A))$
$F(A \lor B) \leftrightarrow F(A) \& F(B)$
$U(A \lor B) \leftrightarrow U(A) \lor U(B)$
$T(\sim A) \quad \leftrightarrow F(A)$
$F(\sim A) \quad \leftrightarrow T(A)$
$U(\sim A) \quad \leftrightarrow U(A)$
$T(\forall xA) \quad \leftrightarrow \forall xT(A)$
$U(\forall xA) \quad \leftrightarrow \exists xU(A)$
$F(\forall yA) \quad \leftrightarrow \forall yP(A) \& \exists yF(A)$
$T(\exists xA) \quad \leftrightarrow \forall xP(A) \& \exists xT(A)$
$U(\exists xA) \quad \leftrightarrow \exists xU(A)$
$F(\exists xA) \quad \leftrightarrow \forall xF(A)$

We shall not offer a full investigation of the Bochvar logic of truth but assign the task to the interested reader.

We have selected Kleene Strong logic to study in detail although any of the three-valued logics from the literature could have been adopted. However, if the techniques for proving consistency which we now study are to be employed, there is one proviso: the logic in question must satisfy a certain criterion of *monotonicity*. This notion can be best explained as follows. Consider the truth tables for Kleene Logic. The logical connectives are *monotone* in the sense that they preserve the partial ordering:

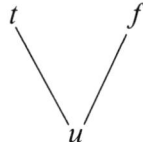

on the truth values. If we write this ordering as $\leqslant$ then it is not difficult to see that $a \leqslant b$ and $a' \leqslant b'$ implies $a \,\&\, a' \leqslant b \,\&\, b'$ where a, a', b, b' are selected from the truth values $\{t, f, u\}$. All the Kleene and Bochvar connectives are, in this sense, monotone.

3.2 Models of KFG

The model theoretic account of **KFG** has its roots in Kripke[1975]. The key idea of the theory insists that the extension of truth should be constructed in a predicative layer-by-layer fashion. One begins with the assignment of truth values to those sentences of L which do not contain the truth predicate. Imagine that one is trying to explain the word "true" to someone who has not grasped its meaning. One might start by introducing the principle encoded in the Tarski biconditionals: to assert that a sentence is true is just to assert the sentence. Consequently, if one can assert

(1)   Snow is white

then one is entitled by the principle to assert:

(2)   'Snow is white' is true

Once the use of the predicate *true* is mastered for these simple sentences, involving no reference to truth, the use of the word true can be extended to sentences which involve the truth predicate itself:

(3)   ' 'Snow is white' is true' is true
(4)   'If 'Snow is white' is true then 'Gulls are white' is true' is true

The intuitive idea behind Kripke's account is based upon this predicative stage-by-stage account of truth. At each stage the notion of truth for sentences of the previous stage is grasped and this, together with the Tarski scheme, enables the notion of truth to be extended to the next stage and so on. Kripke's theory is a formal account of this stage-by-stage analysis of truth. The first step in the theory involves the Kleene semantics for L. This has already been shadowed axiomatically in the theory **KFG** but to develop the Kripke theory we need to spell out the model theoretic account more explicitly.

Let $\mathcal{M} = \langle \mathcal{D}, T, F \rangle$ be a model for L. We employ $\mathcal{M}$ to provide the Kleene semantics for L by defining two semantic relations $\vdash$ (true) and $\dashv$ (false) by simultaneous recursion as follows:

THE STRONG KLEENE TRUTH CONDITIONS FOR L

$\mathcal{M} \vdash_g s = t$ iff $\mathcal{I}[t]_g = \mathcal{I}[s]_g$

$\mathcal{M} \vdash_g T(t)$ iff $T(\mathcal{I}[t]_g) = 1$

$\mathcal{M} \vdash_g F(t)$ iff $F(\mathcal{I}[t]_g) = 1$

$\mathcal{M} \vdash_g A \, \& \, B$ iff $\mathcal{M} \vdash_g A$ and $\mathcal{M} \vdash_g B$

$\mathcal{M} \vdash_g A \vee B$ iff $\mathcal{M} \vdash_g A$ or $\mathcal{M} \vdash_g B$

$\mathcal{M} \vdash_g A \rightarrow B$ iff $\mathcal{M} \dashv_g A$ or $\mathcal{M} \vdash_g B$

$\mathcal{M} \vdash_g \sim A$ iff $\mathcal{M} \dashv_g A$

$\mathcal{M} \vdash_g \forall x A$ iff for all d in D, $\mathcal{M} \vdash_{g(d/x)} A$

$\mathcal{M} \vdash_g \exists x A$ iff for some d in D, $\mathcal{M} \vdash_{g(d/x)} A$

$\mathcal{M} \dashv_g s = t$ iff $\mathcal{I}[t]_g \neq \mathcal{I}[s]_g$

$\mathcal{M} \dashv_g T(t)$ iff $F(\mathcal{I}[t]_g) = 1$

$\mathcal{M} \dashv_g F(t)$ iff $T(\mathcal{I}[t]_g) = 1$

$\mathcal{M} \dashv_g A \, \& \, B$ iff $\mathcal{M} \dashv_g A$ or $\mathcal{M} \dashv_g B$

$\mathcal{M} \dashv_g A \vee B$ iff $\mathcal{M} \dashv_g A$ and $\mathcal{M} \dashv_g B$

$\mathcal{M} \dashv_g A \rightarrow B$ iff $\mathcal{M} \vdash_g A$ and $\mathcal{M} \dashv_g B$

$\mathcal{M} \dashv_g \sim A$ iff $\mathcal{M} \vdash_g A$

$\mathcal{M} \dashv_g \forall x A$ iff for some d in D, $\mathcal{M} \dashv_{g(d/x)} A$

$\mathcal{M} \dashv_g \exists x A$ iff for all d in D, $\mathcal{M} \dashv_{g(d/x)} A$

This semantic account is nothing more than the model-theoretic version of **KFG**. Observe that we do not always have $\mathcal{M} \vdash_g A$ or $\mathcal{M} \dashv_g A$: not every wff is rendered true or false by this semantics since the atomic assertions involving the truth and falsity predicates may not be rendered either true or false. More specifically, if $T(\mathcal{I}[t]_g) = 0$ and $F(\mathcal{I}[t]_g) = 0$ then t is undecided and consequently where $A = T(t)$ we do not have $\mathcal{M} \vdash_g A$ or $\mathcal{M} \dashv_g A$. The semantics thus admits *truth value gaps*. Of course, the original extensions of truth and falsity also admit *gaps* in the sense that we may have both $T(d) = 0$ and $F(d) = 0$. One can think of matters in a slightly different way by packing the information contained in T and F into one predicate $S: D \rightarrow \{0, 1, u\}$ where

$$S(d) = \begin{cases} 1 \text{ if } T(d) = 1 \\ 0 \text{ if } F(d) = 1 \\ u \text{ if otherwise} \end{cases}$$

where the value u represents the undefined value. In arbitrary models the extensions of truth and falsity (T and F) are completely unconstrained and certainly do not satisfy the axioms of **KFG**. The relations $\vdash$ and $\dashv$ are somewhat better candidates for the extensions in that they assign the Kleene theory of truth and falsity correctly to those sentences of L which do not contain the truth or falsity predicates. This suggests that we revise the extensions of truth and falsity to capture this insight. This is the formal counterpart of the layer-by-layer build-up of the extension of truth which informed the informal account.

DEFINITION 3.0
Let $\mathcal{M} = \langle \mathcal{D}, T, F \rangle$ be a model for L. Define $\mathcal{M}' = \langle \mathcal{D}, T', F' \rangle$, the **Kleene revision of** $\mathcal{M}$, by

$$T'(\mathcal{I}[A]_g) = 1 \quad \text{iff} \quad \mathcal{M} \vdash_g A$$
$$F'(\mathcal{I}[A]_g) = 1 \quad \text{iff} \quad \mathcal{M} \dashv_g A$$

On those elements of D which are not elements of CODE, T and F do not change. This definition is legitimate since the coding function enjoys the aforementioned *independence* property.

Unfortunately, although this revision gets the truth-conditions for the first-stage sentences correct, it does little in general for sentences which involve two levels of the truth-predicate. To reach these we have to repeat the process of revision. Firstly, however, we have to check that once a sentence has been definitely marked as true (or false) further revisions will not affect matters. This is formally reflected in the following concept and theorem.

DEFINITION 3.1
Let $\mathcal{M} = \langle \mathcal{D}, T, F \rangle$ and $\mathcal{N} = \langle \mathcal{D}, \hat{T}, \hat{F} \rangle$ be models for L.
Define $\mathcal{M} \subseteq \mathcal{N}$ iff $\forall d \in D(T(d) = 1 \to \hat{T}(d) = 1 \,\&\, F(d) = 1 \to \hat{F}(d) = 1)$.
Define $\mathcal{M} \equiv \mathcal{N}$ iff $\mathcal{M} \subseteq \mathcal{N}$ and $\mathcal{N} \subseteq \mathcal{M}$.

In other words, one model is an *extension* of a second when the predicates of truth and falsity of the first are extensions of (or equal to) those of the second.

Kleene revision reflects the intuitive idea of building up the extensions of truth and falsity in a predicative stage-by-stage fashion. Given that we have mastered the extensions of truth and falsity at a certain stage, we are free to move to a new level and extend our notions to sentences of this new stage, safe in the knowledge that a sentence once marked as true or false remains so. At least, this is the ideal, which is fortunately justified by the following result.

THEOREM 3.2
Let $\mathcal{M}$ and $\mathcal{N}$ be two models of L. Then $\mathcal{M} \subseteq \mathcal{N}$ implies $\mathcal{M}' \subseteq \mathcal{N}'$.

PROOF By induction on wff we show that $\mathcal{M} \vdash_g A$ implies $\mathcal{N} \vdash_g A$ and $\mathcal{M} \dashv_g A$ implies $\mathcal{N} \dashv_g A$. The atomic cases both follow directly from the assumption. For the induction step we illustrate the various cases with negation and disjunction. For negation, observe that $\mathcal{M} \vdash_g \sim A$ iff $\mathcal{M} \dashv_g A$ implies $\mathcal{N} \dashv_g A$ (by induction) iff $\mathcal{N} \vdash_g \sim A$. Similarly, $\mathcal{M} \dashv_g \sim A$ iff $\mathcal{M} \vdash_g A$ implies $\mathcal{N} \vdash_g A$ (induction) iff $\mathcal{N} \dashv_g \sim A$. For conjunction, $\mathcal{M} \vdash_g A \& B$ iff $\mathcal{M} \vdash_g A$ and $\mathcal{M} \vdash_g B$, which by induction implies $\mathcal{N} \vdash_g A$ and $\mathcal{N} \vdash_g B$, which yields $\mathcal{N} \vdash_g A \& B$. The argument for the $\mathcal{M} \dashv_g A \& B$ case is similar. ●

In general, one revision will not be sufficient. We need the revision process to reflect the arbitrary depth of possible embeddings of the truth and falsity predicates. This stage-by-stage process is captured as follows. Using this basic step of Kleene revision we can define an ordinal sequence of truth and falsity predicates $T(\alpha)$, $F(\alpha)$ and models $\mathcal{M}(\alpha)$ for $\alpha \geqslant 0$. In what follows $\mathbb{0}$ is the identity function on D with value 0.

$$T(0) = \mathbb{0}$$
$$F(0) = \mathbb{0}$$
$$T(\alpha + 1) = T(\alpha)'$$
$$F(\alpha + 1) = F(\alpha)'$$
$$T(\delta)(d) = 1 \text{ iff } (\exists \alpha < \delta)(\forall \beta)(\alpha \leqslant \beta < \delta)(T(\beta)(d) = 1),$$
$$\text{for limit ordinal } \delta$$
$$F(\delta)(d) = 1 \text{ iff } (\exists \alpha < \delta)(\forall \beta)(\alpha \leqslant \beta < \delta)(F(\beta)(d) = 1),$$
$$\text{for limit ordinal } \delta$$

The form of this definition is quite general and will be employed again in the next chapter. Notice, however, that the monotonicity of Kleene revision (3.2) implies that the definitions of truth and falsity at limit ordinals take on a more perspicuous form:

$$d \in T(\delta) \quad \text{iff } d \in \bigcup_{\alpha < \delta} T(\alpha)$$

$$d \in F(\delta) \quad \text{iff } d \in \bigcup_{\alpha < \delta} F(\alpha)$$

where here the extensions of truth and falsity are considered as sets. The fact that we started with $\mathbb{0}$ is compensated for by the truth conditions for equality: after one revision the extensions of truth and falsity will be nonempty.

The various finite revisions in the sequence reflect the various stages of the informal account and the degree to which the truth predicate is embedded in the sentences of L. Indeed, the various stages reflect ever better approximations to the extensions of truth and falsity and these stages, as we now see, culminate in a model for **KFG**. The following *fixpoint* theorem facilitates the construction of models for **KFG**.

THEOREM 3.3
There is a model $\mathcal{M}^* = \langle \mathcal{D}, T^*, F^* \rangle$ such that $\mathcal{M}^* \equiv (\mathcal{M}^*)'$.

PROOF We prove by induction on ordinals that for all α,

$$\mathcal{M}(\alpha) \subseteq \mathcal{M}(\alpha + 1)$$

This holds for $\alpha = 0$ by definition. Moreover, if $\mathcal{M}(\alpha) \subseteq \mathcal{M}(\alpha + 1)$ then, by monotonicity,

$$\mathcal{M}(\alpha + 1) \subseteq \mathcal{M}(\alpha + 2)$$

Furthermore, for limit ordinals λ, we have

$$\mathcal{M}(\lambda) = \bigcup_{\alpha < \lambda} \mathcal{M}(\alpha) = \bigcup_{\alpha < \lambda} (\mathcal{M}(\alpha))' \subseteq \left(\bigcup_{\alpha < \lambda} \mathcal{M}(\alpha) \right)' = \mathcal{M}(\lambda)'$$

It follows that $\alpha < \beta$ implies that $\mathcal{M}(\alpha) \subseteq \mathcal{M}(\beta)$. Now suppose that $\mathcal{M}(\alpha) \subset \mathcal{M}(\alpha)'$ and $\mathcal{M}(\alpha) \neq \mathcal{M}(\alpha)'$, for each α. Then choose some $x(\alpha) \in (\mathcal{M}(\alpha)' - \mathcal{M}(\alpha))$, for each α less than $|\mathcal{M}|^*$ (the least cardinal number greater than the cardinality of $\mathcal{M}$ (the domain of the model)). Put $X = \{x(\alpha) : \alpha < |\mathcal{M}|^*\}$. This is a subset of $\mathcal{M}$ of cardinality $|\mathcal{M}|^*$. This is a contradiction and so our assumption that $\mathcal{M}(\alpha) \neq \mathcal{M}(\alpha)'$ for all α is false. Hence, there must be a least α such that $\mathcal{M}(\alpha) \equiv \mathcal{M}(\alpha)'$. Let $\mathcal{M}^*$ be this $\mathcal{M}(\alpha)$. ●

THEOREM 3.4
The model $\mathcal{M}^* = \langle \mathcal{D}, T^*, F^* \rangle$ is a model of the theory **KFG**.

PROOF The soundness of the axioms is automatic from the Kleene Truth conditions and the facts:

$$T^*(\mathcal{I}[A]_g) = 1 \quad \text{iff} \quad \mathcal{M}^* \vDash_g A$$
$$F^*(\mathcal{I}[A]_g) = 1 \quad \text{iff} \quad \mathcal{M}^* \nvDash_g A \quad ●$$

Observe that we do not, as a consequence of the above, have the agreement of the Kleene and classical truth conditions, i.e. the following are generally false:

$$\mathcal{M}^* \vDash_g A \quad \text{iff} \quad \mathcal{M}^* \vDash_g A$$
$$\mathcal{M}^* \nvDash_g A \quad \text{iff} \quad \mathcal{M}^* \vDash_g {\sim} A$$

Put slightly differently, we have for classical semantics:

$$\mathcal{M} \vDash_g \sim A \quad \text{iff not } (\mathcal{M} \vDash_g A) \qquad \text{Classical}$$

whereas for Kleene semantics we only have:

$$\mathcal{M} \dashv_g A \quad \text{iff } \mathcal{M} \vDash_g \sim A \qquad \text{Kleene}$$

Consequently, and fortunately, the Tarski biconditionals do not follow.

We now turn our attention to the notion of proposition. If you recall, propositions are those wff which are true or false. The notion is captured model-theoretically by Kripke's notion of *grounded* wff.

DEFINITION 3.5
Let $\mathcal{M}^*$ be a fixpoint model. A wff A is *positively grounded* iff $\mathcal{M}^* \vdash A$; it is *negatively grounded* iff $\mathcal{M}^* \dashv A$; it is *grounded* iff it is positively or negatively grounded.

Alternatively, it is positively grounded iff $T^*(\mathcal{I}[A]_g) = 1$ and negatively grounded if $F^*(\mathcal{I}[A]_g) = 1$.

This completes our account of the model-theory of **KFG**. In Kripke's theory the language L is itself provided with a Kleene semantics and so the actual logic is three-valued. Our account is essentially Kripke's but with the modification that L is interpreted classically in the model $\mathcal{M}^*$. The Kripke theory is a very simple and appealing one. The axiomatic theory **KFG**, which goes with it, is also elegant and simple although, as we shall see, it does have its oddities.

3.3   A Reformulation

We now return to the theory **KFG** and examine a little of its infrastructure. A somewhat different formulation of the theory was supplied by Feferman [1984]. In the latter formulation the theory is not explicitly given in terms of the Kleene truth-conditions but rather in terms of the schemata:

    C1 $T(A) \leftrightarrow A^+$
    C2 $F(A) \leftrightarrow A^-$

where A^+ and A^- are in a sense *approximations* to A and $\sim A$, respectively. In essence, these approximations are the result of pushing all negations into atomic position and replacing all such negations by *internal negations*. The precise definition follows.

DEFINITION 3.6
Let A be any wff of L. Then A^+ and A^- are defined by simultaneous recursion:

(i)   If A is $t = s$ then $A^+ = A$ and $A^- = t \neq s$
(ii)   If A is $T(t)$ then $A^+ = A$ and $A^- = F(t)$
       If A is $F(t)$ then $A^+ = A$ and $A^- = T(t)$
(iii) If A is $\sim B$ then $A^+ = B^-$ and $A^- = B^+$
(iv)   If A is $B \& C$ then $A^+ = B^+ \& C^+$ and $A^- = B^- \vee C^-$
(v)   If A is $B \vee C$ then $A^+ = B^+ \vee C^+$ and $A^- = B^- \& C^-$
(vi)   If A is $\forall xB$ then $A^+ = \forall xB^+$ and $A^- = \exists xB^-$
(vii) If A is $\exists xB$ then $A^+ = \exists xB^+$ and $A^- = \forall xB^-$
(viii) If A is $B \to C$ then $A^+ = B^{\ddot{}} \vee C^+$ and $A^- = B^+ \& C^-$

The following captures the elementary properties of these approximations.

LEMMA 3.7(i)
For each wff A we have:

(i) $A^+ \to A$
(ii) $A^- \to \sim A$
(iii) $(A \leftrightarrow A^+) \to (T(A) \leftrightarrow A)$
(iv) $(\sim A \leftrightarrow A^-) \to (F(A) \leftrightarrow \sim A)$
(v) $(A^+ \leftrightarrow A \& A^- \leftrightarrow \sim A) \leftrightarrow P(A)$

PROOF   The proof for (i) and (ii) is routine and is established by simultaneous induction. We illustrate with negation: $(\sim A)^+ = A^-$. We have $A^- \to \sim A$ (by induction) and so $(\sim A)^+ \to \sim A$. Next observe $(\sim A)^- = A^+$. We have $A^+ \to A$ (by induction). Since $A \leftrightarrow \sim \sim A$ we have $(\sim A)^- \to \sim(\sim A)$. The third, fourth and fifth parts are immediate.

Part (v) is interesting in that it provides a necessary and sufficient criteria for being a proposition. With these preliminary points covered we come to the main point of this section: the next result informs us that the present theory is nothing more than a reformulation of KFG.

THEOREM 3.8
The theory KFG is equivalent to DIS + (C1 – C2).

PROOF   It is obvious that each of the axioms A1–A6 and F1–F12 follow from C1 and C2. We illustrate with F9:

$$T(\forall xA) \leftrightarrow (\forall xA)^+ \leftrightarrow \forall xA^+ \leftrightarrow \forall xT(A)$$

The converse direction is by induction on wff. Use A1−A6 for the atomic cases and F1−F12 for the induction clauses. ●

This version of the theory slightly hides its three-valued origins and the theory is best summarized by saying that T(A) means that A is Kleene-true and F(A) that A is Kleene-false.

3.4 Truth, Falsity and Propositions

The notion of "groundedness" which arises from the model theory is reflected by the notion of "proposition". Giving up the classical logic of truth **CL** means giving up bivalence so not everything is a proposition. Indeed, it is easy to see that **KFG** + PB ≡ **CL**. Our next task is to investigate the properties and truth conditions for this notion of proposition.

THEOREM 3.9

In the theory **KFG** the notion of proposition satisfies the following:

    (i) $P(A) \& P(B) \rightarrow P(A \& B)$
    (ii) $P(A) \& P(B) \rightarrow P(A \vee B)$
    (iii) $P(A) \& P(B) \rightarrow P(A \rightarrow B)$
    (iv) $P(A) \rightarrow P(\sim A)$
    (v) $\forall x P(A) \rightarrow P(\forall x A)$
    (vi) $\forall x P(A) \rightarrow P(\exists x A)$
    (vii) $P(s = t)$

  PROOF   To illustrate we consider (i). Suppose that

    $(T(A) \vee F(A)) \& (T(B) \vee F(B))$

Then by the Propositional Calculus it follows that

    $(T(A) \& T(B)) \vee F(A) \vee F(B)$

By **KFG** axioms for conjunction we obtain: $T(A \& B) \vee F(A \& B)$, as required.
●

The condition for implication can be strengthened to

    (iii′) $P(A) \& (T(A) \rightarrow P(B)) \rightarrow P(A \rightarrow B)$

This reflects the idea that the propositional nature of B is *dependent* upon the truth of A. These axioms for propositionhood are the minimum one might expect for a theory of propositions in that they only guarantee closure of the notion under the logical connectives and quantifiers. We next examine their truth conditions: on those wff which are propositions, we obtain the standard Tarski truth conditions. We leave the reader to prove the following.

THEOREM 3.10
In **KFG** the notions of truth and proposition satisfy the following:

    (i) $P(A) \mathbin{\&} P(B) \rightarrow (T(A \mathbin{\&} B) \leftrightarrow T(A) \mathbin{\&} T(B))$

    (ii) $P(A) \mathbin{\&} P(B) \rightarrow (T(A \vee B) \leftrightarrow T(A) \vee T(B))$

    (iii) $P(A) \mathbin{\&} P(B) \rightarrow (T(A \rightarrow B) \leftrightarrow (T(A) \rightarrow T(B)))$

    (iv) $P(A) \rightarrow (T(\sim A) \leftrightarrow \sim T(A))$

    (v) $\forall x P(A) \rightarrow (T(\forall x A) \leftrightarrow \forall x T(A))$

    (vi) $\forall x P(A) \rightarrow (T(\exists x A) \leftrightarrow \exists x T(A))$

    (vii) $T(s = t) \leftrightarrow s = t$

Once again (iii) can be strengthened to

    (iii′) $(P(A) \mathbin{\&} (T(A) \rightarrow P(B))) \rightarrow (T(A \rightarrow B) \leftrightarrow (T(A) \rightarrow T(B)))$

This theory does not assign the standard Tarski truth conditions to all the wff but only those which are provably propositions. As we have said before, these axioms reflect the minimal conditions one would expect of any theory of truth. We shall refer to the theory (given by 3.9 and 3.10 including the modifications to the axioms of implication) as **TP**. We shall actually explore this theory in more detail in chapter 5 where it forms the basis for Peter Aczel's theory of *Frege Structures*.

    However, the theory **KFG** supports a little more. More precisely, in addition to 3.9 and 3.10 we have the following equivalences:

THEOREM 3.11
The following are provable in **KFG**:

    (i) $P(A) \leftrightarrow P(T(A))$

    (ii) $P(A) \leftrightarrow T(P(A))$

    PROOF   Both are straightforward and we assign them as exercises for the reader.

Moreover, we can derive certain rules of inference which govern the truth predicate. The following is a natural deduction formulation of a *logic of truth*.

Conjunction
$$\frac{T(A) \qquad T(B)}{T(A \;\&\; B)} \qquad\qquad \frac{T(A \;\&\; B)}{T(A)} \qquad \frac{T(A \;\&\; B)}{T(B)}$$

Implication

$$[T(A)]$$
$$\vdots$$
$$\frac{T(B) \qquad P(A)}{T(A \to B)} \qquad\qquad \frac{T(A) \qquad T(A \to B)}{T(B)}$$

Disjunction

$$\frac{T(A)}{T(A \lor B)} \qquad\qquad \frac{T(A \lor B) \qquad T(C) \qquad T(C)}{T(C)}$$
with discharged assumptions $[T(A)] \quad [T(B)]$ above the two $T(C)$

$$\frac{T(B)}{T(A \lor B)}$$

Negation

$$[T(A)]$$
$$\vdots$$
$$\frac{\bot \qquad P(A)}{T(\sim A)} \qquad\qquad \frac{T(A) \qquad T(\sim A)}{\bot}$$

Classical Negation

$$[T(\sim A)]$$
$$\vdots$$
$$\frac{\bot \qquad P(A)}{T(A)}$$

Universal Quantification
$$\frac{T(A[c])}{T(\forall x A)} \qquad\qquad \frac{T(\forall x A)}{T(A[t])}$$

Existential Quantification
$$\frac{T(A[t])}{T(\exists x A)} \qquad\qquad \frac{T(\exists x A) \qquad T(C)}{T(C)}$$
with discharged assumption $[T(A[c])]$ above $T(C)$

We assume the normal side conditions on the quantifier rules for existential elimination and universal introduction, i.e. in the universal introduction rule c is not to occur in any assumption on which $A[c]$ depends and in the existential elimination rule c must not occur in $T(C)$ or in any assumption on which $T(A[c])$ depends. Observe that there are side conditions on propositions in the implication introduction rule whereas there are no such conditions for disjunction or existential elimination. This might be considered somewhat strange. The reader should compare these logics with the combinatory logics which mention propositions in Hindley and Seldin.

THEOREM 3.12
The above natural deduction system is derivable in **KFG**.

PROOF We illustrate with implication introduction. Assume P(A) and T(A). By the assumption of the rule we obtain T(B). Hence P(B) and T(A) → T(B) — discharging the assumption T(A). By the theory **TP** we obtain T(A → B), as required. ●

The reader should by now have obtained some intuitive grasp of this logic and indeed some understanding of its limitations. We now explore its structure just a little more with a view to pointing out some oddities.

3.5 KFG as a Modal Theory

To prepare the route to the next theory we investigate some more of the consequences of **KFG** and in particular the derivability of certain modal principles.

THEOREM 3.13
The following are provable in **KFG**.

R	$T(A) \to A$
Tr	$T(A) \to T(T(A))$
IP	$T(A \to B) \to (T(A) \to T(B))$
BAR	$\forall x T(A) \to T(\forall x A)$

PROOF We illustrate with IP. By definition, $T(A \to B) \leftrightarrow (A^- \vee B^+)$. Assume T(A); by our reformulation we have A^+. This implies $\sim A^-$ and so B^+, i.e. T(B). ●

These are the characteristic axioms of **S4** modal logic. As regards the **S5** axiom, which in the present setting takes the form $C(A) \to T(C(A))$ (where C(A) abbreviates $\sim T(\sim A)$), note that the following is provable:

$$(C(A) \to T(C(A))) \to P(A)$$

As a consequence, **S5** is not derivable: its truth renders everything a proposition and hence the theory inconsistent. However, there are further modal-style axioms which are provable.

THEOREM 3.14
The following are (easily) provable in **KFG**:

S	$T(T(A) \rightarrow A) \leftrightarrow P(A)$
IPT	$T(T(A) \rightarrow T(B)) \leftrightarrow T(A \rightarrow B)$
N	$T(A) \rightarrow T(C(A))$
L	$T(C(A)) \rightarrow T(A)$

With these axioms in place it appears that **KFG** supports a distinctive modal theory of truth but one important factor is missing, namely a rule of necessitation:

$$\textbf{KFG} \vdash A \text{ implies } \textbf{KFG} \vdash T(A)$$

This rule is not provable in **KFG**. Indeed, its addition renders the theory inconsistent: to see this, observe $\textbf{KFG} \vdash A \vee \sim A$; hence by the rule $\textbf{KFG} \vdash T(A \vee \sim A)$ and so by the axiom for disjunction we have $T(A) \vee F(A)$, i.e. everything will be a proposition. Indeed this follows from a weak rule of necessitation which only allows the derivation of $T(A)$ when A is provable in classical logic. Although the logic has the appearance of a modal theory, it is not very interesting given the lack of any obvious rule of necessitation. Indeed, the schemata C1 and C2 under the above rule would lead to

$$T(T(A) \leftrightarrow A^{+})$$
$$T(F(A) \leftrightarrow A^{-})$$

but the truth of either again yields that everything is a proposition. This is rather unfortunate since one would like the assertion of the truth of the schema of the theory to be true. We shortly turn to a theory which looks more promising as a modal theory.

3.6 The Strengthened Liar

No account of the fixpoint theory of truth would be complete without some discussion of the *strengthened liar*. The usual formulation is couched in terms of the following sentence:

S: If S is a proposition then S is false.

In other words, S says of itself that if it is a proposition then it is false. In the formal language of the theory this would be rendered as

$$S \leftrightarrow (P(S) \rightarrow F(S))$$

Such a sentence has its existence guaranteed by the diagonalization lemma. The theory **KFG** is consistent and so S cannot yield an inconsistency. Nevertheless, it will be somewhat instructive to see what the theory says about S. The intuitive result we expect is that it deals with S in much the same way as the ordinary *liar*: it is banished from being a proposition. Is S a proposition? Suppose it is, i.e. P(S). Then we obtain F(S) and consequently ~S. But then ~(P(S) → F(S)) and since P(S) we must have ~F(S) — a contradiction. So formally S cannot be a proposition and our intuitions are satisfied.

This ends our exposition of the three-valued approach. It has, as one might expect, some good points and some bad ones. Firstly, it has an elegant and intuitively appealing underlying semantic theory of revision. It is founded upon the predicative notion of truth built into any such process of revision. Secondly, the theory is easy to state: either of the two formulations are simple and elegant. Subsequently, the logic is easy to work with. The negative aspects relate to its three-valued origins. It still seems philosophically odd that a theory of truth for classical logic should appeal to non-classical semantics in such a crucial way. We now turn to an approach which has its origins not in any three-valued truth conditions but rather in the classical truth conditions themselves. It thus builds upon the insights of Kripke's predicative process of revision but employs classical semantics rather than the Kleene counterpart.

4   Stable Truth

The theory **KFG** is a logic of truth based upon three-valued logic. As a consequence the principle of bivalence

PB $T(A) \vee F(A)$

has to be sacrificed. This is an inevitable consequence of maintaining classical logic, but what is not is the other consequence of the three-valued approach, namely the rejection of the principle (LT) which demands that every logical truth be true.

LT   If A is a logical truth then $T(A)$.

In **KFG** the principle has to go since, as we previously indicated, it is inconsistent with the theory. This does seem to be at variance with the classical conception of truth: surely all classical truths should be true. In this chapter we investigate a logic of truth which upholds LT. The theory arises from a revision process which builds upon the classical truth conditions.

The account of revision presented in chapter 3 employs Kleene logic in that the model-theoretic account of **KFG** is based upon a theory of revision in which the extensions of truth and falsity are determined by the Kleene truth conditions. This process of revision departs from the classical conception of truth in that some form of (monotone) non-classical semantics is employed at the revision step. Certain reasons might be marshalled to defend the use of monotone logics in the revision process but there is little doubt that the natural (or perhaps naive) logic of truth is embodied in the classical truth conditions which are definitely not monotone. This does, however, raise an important question: what happens if we use the classical semantics at the point of revision? In this chapter we address this question and develop the underlying logic of truth. On the face of it things should go badly wrong since we seem to be building in the very conditions which we have discarded in order to avoid contradiction. In fact, things go rather well and we obtain an elegant and pleasing theory.

4.1 The Gupta–Herzberger Semantic Theory

The Gupta–Herzberger semantic theory of truth adopts a theory of revision built upon classical semantics. The intuitions which drive the theory are similar to those which underpin the Kripke stages; the difference is located in the definition of revision. Instead of employing the *Kleene revision* we adopt the more *naive* approach and use the classical Tarski truth conditions.

DEFINITION 4.0
Let $\mathcal{M} = \langle \mathcal{D}, T, F \rangle$ be a model for L.
  Define $\mathcal{M}' = \langle \mathcal{D}, T', F' \rangle$ the *Tarski revision of* $\mathcal{M}$ by

$$T'(\mathscr{I}[A]_g) = 1 \quad \text{iff} \quad \mathcal{M} \vDash_g A$$
$$F'(\mathscr{I}[A]_g) = 1 \quad \text{iff} \quad \mathcal{M} \vDash_g \sim A$$

Once again on those elements of D which are not elements of CODE, T and F do not change.

Starting from any model and employing this basic step of revision we can define a sequence of truth and falsity predicates $T(\alpha)$, $F(\alpha)$ for $\alpha \geq 0$, in much the same way as Kleene revision.

(i) $T(0) = T$
                $F(0) = F$
(ii) $T(\alpha + 1) = T(\alpha)'$
                $F(\alpha + 1) = F(\alpha)'$
(iii)   For limit ordinal δ define:
                $T(\delta)(d) = 1 \quad \text{iff} \quad \exists\alpha(\alpha < \delta)\forall\beta(\alpha \leq \beta < \delta)(T(\beta)(d) = 1)$
                $F(\delta)(d) = 1 \quad \text{iff} \quad \exists\alpha(\alpha < \delta)\forall\beta(\alpha \leq \beta < \delta)(F(\beta)(d) = 1)$

Actually there are many options for the definition of the truth predicate at limit ordinals and different choices lead to different semantic theories of truth. The articles by Gupta and Herzberger contain indications of the different choices available, but we shall not pursue this here. We have chosen the one which appears most susceptible to an elegant axiomatization. We recommend that the reader examine these other options and investigate the logics along the lines developed here.

    Observe that the starting model is left arbitrary with respect to truth and falsity. This should be seen in contrast to the Kleene revision where we started with the empty truth and falsity predicates (or more generally, a model $\mathcal{M}$ such that $\mathcal{M} \subseteq \mathcal{M}'$). As we shall see, in the case of Tarski revision, the nature of the starting model can be largely ignored. As the process of revision proceeds, the role of the initial model is gradually eroded.

The difference between Tarski revision and Kleene revision concerns the nature of the revision step. In the case of the latter the process is monotone but in the former case it is not. For example, the *liar* sentence switches truth value at every revision. To see this, suppose that A is the version of the liar sentence which satisfies $A \leftrightarrow T(\sim A)$. If A is true at the initial model $\mathcal{M}$ then A cannot be true at $\mathcal{M}'$, the revision of $\mathcal{M}$. Suppose A is true at $\mathcal{M}'$, then since $A \leftrightarrow T(\sim A)$, we have $T(\sim A)$ is true at $\mathcal{M}'$. Hence $\sim A$ is true at $\mathcal{M}$. This is a contradiction so $\sim A$ is true at $\mathcal{M}'$. If, on the other hand, $\sim A$ is true at $\mathcal{M}$, then $T(\sim A)$ will be true at $\mathcal{M}'$ and consequently by $A \leftrightarrow T(\sim A)$, A will be true at $\mathcal{M}'$. Despite this undisciplined behaviour, there is a counterpart to groundedness.

DEFINITION 4.1
An element d in D is *positively stable* iff $\exists \alpha \forall \beta \geq \alpha (T(\beta)(d) = 1)$.
  It is *negatively stable* iff $\exists \alpha \forall \beta \geq \alpha (F(\beta)(d) = 1)$.
  An element d of D is *stable* iff d is positively or negatively stable. We say that d is *positively stable from an ordinal* α [*negatively stable from* α] iff

$$\forall \beta \geq \alpha (T(\beta)(d) = 1) \quad [\forall \beta \geq \alpha (F(\beta)(d) = 1)]$$

In the case of Kleene revision we were able to locate a model (the least fixpoint model) in which the grounded elements are delineated. The non-monotone behaviour of the Tarski revision does not facilitate the construction of fixpoint models. There is, however, an analogous notion which isolates all the stable elements.

DEFINITION 4.2
An ordinal σ is a *stabilization ordinal* iff

(i)   For each d in D, d is positively stable iff $T(\sigma)(d) = 1$.
      For each d in D, d is negatively stable iff $F(\sigma)(d) = 1$.
(ii) For each d in D, d is positively (negatively) stable implies that d is positively (negatively) stable from σ.

Stabilization ordinals characterize the stable objects exactly. Part (i) of the above informs us that something is positively (negatively) stable iff it is in the extension of truth (falsity) at stabilization ordinals, and part (ii) says that if something is (positively or negatively) stable it is so from such an ordinal. It is all very well postulating such ordinals but do such models exist?

THEOREM 4.3
There exists a stabilization ordinal.   (Herzberger)

Indeed, the stabilization ordinals occur with a fixed periodicity. The actual mathematics of the stabilization process is not devoid of interest. However, it should not detain us too much since we are primarily concerned with the logic of truth which is characterized by the stabilization process. It is worth pointing out, however, that this (semi-inductive) process is perfectly general and applies to any family of sets B (which are subsets of some set U) together with a *jump* operation $J: B \to B$. Given such a framework, we can define an ordinal sequence of sets selected from B by

$$f(\alpha + 1) = J(f(\alpha))$$
$$f(\lambda) = \{u \in U : \exists \alpha < \lambda \forall \beta (\alpha \leqslant \beta < \lambda)(u \in f(\beta))\} \qquad \text{for limit ordinal } \lambda$$

Under such general conditions stabilization ordinals exist. The process of revision takes on the following general form. First a *saturation point* is reached. This is an ordinal α such that for every x in the range of f, there is an ordinal $\beta < \alpha$ such that $f(\beta) = x$. In other words, the range of f has been exhausted by the time α is reached. Once a saturation point is located the stages are all repetitions of previous stages, and the process cycles. However, the first cycle is not necessarily the last. The process goes through a sequence of cycles of increasing length until a *grand loop* is reached. After this point the process cycles permanently through the stages in the loop. For more details see Herzberger [1982], Gupta [1982] and Visser [1984]. This is only a brief account of the Gupta-Herzberger theory of truth. It is sufficient for our purposes but the papers referenced contain more details.

4.2 Logics of Stable Truth

Our main interest throughout this book concerns the actual logics of truth and modality. In the present setting of the stabilization process there are two concepts of validity which are, prima facie, of interest. The first is generated by that class of sentences true at all stabilization models while the second corresponds to that class which are stably true. These two classes are clearly related in that T(A) is in the first class iff A is in the second. These classes are captured precisely by the following definitions.

DEFINITION 4.4
A wff A is *safe* iff A is valid at every stabilization ordinal for every starting model. A is *stably true* iff T(A) is safe and A is *stably false* iff F(A) is safe.

Our goal is to explore the logics of safeness and stability. What form might such logics take? In addressing this question observe that the

Gupta–Herzberger semantics for the truth predicate has a *modal* flavour to it: for a wff A to be stably true A must be true in all models after some point in the revision process. Moreover, at stabilization ordinals T(A) means that A is stably true and F(A) that A is stably false. In the rest of this chapter we shall explore this *modal* interpretation of the truth predicate and consider various modal logics all consistent with this stability interpretation. The modal logics we obtain are somewhat nonstandard. We shall not give the strongest logics immediately. Our strategy will be to explore the space of all the logics which are sound under the revision interpretation beginning with the weakest logics.

4.3   A Weak Modal Logic of Truth

Actually, the weakest system we shall consider is not really capable of being interpreted as a logic of necessity and possibility; instead it is a weak deontic logic. Nevertheless, the theory has some intrinsic interest. Throughout we shall adopt the principle $F(A) \leftrightarrow T(\sim A)$.

THE THEORY M

DIS $T(A) \rightarrow C(A)$ where $C(A) =_{\text{def}} \sim F(A)$
IP $T(A \rightarrow B) \rightarrow (T(A) \rightarrow T(B))$
BAR $\forall x T(A) \rightarrow T(\forall x A)$
NEC   If $\textbf{LC} \vdash A$ then $T(A)$ is a theorem of **M**

This is a very weak modal theory. In particular, it has a weakened rule of necessitation. One can only conclude that T(A) if A is derivable from the underlying theory of the Lambda Calculus **LC**. The axiom IP is self-explanatory. The axiom BAR is essentially the Barcan formulae. The axiom DIS, familiar from deontic logic, prevents both T(A) and F(A) being simultaneously true.

We first establish that all the theorems of **M** are true at any stabilization model.

THEOREM 4.5
If $\textbf{M} \vdash A$ then A is safe.

PROOF   The axiom IP follows because Modus Ponens preserves stability: if T(A) and $T(A \rightarrow B)$ are true at a stabilization ordinal then A and $A \rightarrow B$ will be stably true and so B will. Hence T(B) will be true at this stabilization ordinal. Next consider the Barcan formulae. This follows because the domain of individuals is fixed throughout the revision process. The inference rule of weak

49

necessitation NEC preserves soundness because any wff provable in the underlying theory of the Lambda Calculus will be true in all models and will thus be stably true. DIS is true at any stabilization ordinal because T(A) means that A will be true from the stabilization point onwards and so there is no possibility of A being false let alone stably false. ●

This is a relative consistency result for the logic **M** since all its theorems are true in some model; in this case they are true at any stabilization model.

This theory already supports the intuition that all first-order logical truths should be true. Intuitively, if these are not, it is difficult to imagine what is. Prompted by this we might enquire how much of **KFG** remains intact.

THEOREM 4.6
The following are derivable in **M**.

  (i) $T(A \& B) \leftrightarrow T(A) \& T(B)$
  (ii) $F(A) \vee F(B) \rightarrow F(A \& B)$
  (iii) $T(A) \vee T(B) \rightarrow T(A \vee B)$
  (iv) $F(A \vee B) \leftrightarrow F(A) \& F(B)$
  (v) $T(\sim A) \leftrightarrow F(A)$
  (vi) $F(\sim A) \leftrightarrow T(A)$
  (vii) $T(B) \vee F(A) \rightarrow T(A \rightarrow B)$
  (viii) $F(A \rightarrow B) \leftrightarrow T(A) \& F(B)$
  (ix) $T(\forall x A) \leftrightarrow \forall x T(A)$
  (x) $\exists x F(A) \rightarrow F(\forall x A)$
  (xi) $\exists x T(A) \rightarrow T(\exists x A)$
  (xii) $F(\exists x A) \leftrightarrow \forall x F(A)$

  PROOF   These are all quite straightforward and follow from the appropriate classical truths and judicious applications of IP, BAR and NEC. We illustrate with (i). From left to right we employ the tautology $A \& B \rightarrow A$, NEC and IP; from right to left we employ the tautology $A \rightarrow (B \rightarrow A \& B)$, NEC and two applications of IP. ●

The converse directions of (ii), (iii), (vii), (x), (xi) are not generally derivable. In fact, the theory which results from the inclusion of these converse directions gives precisely the strong Kleene truth conditions for T and F, i.e. the theory **KFG**. To keep necessitation, and consequently the fact that all logical truths are true, we must give up these principles.

We have already located the major difference between **KFG** and the logic of stable truth: the latter supports LT but the former does not. In this sense **KFG** is not classical whereas the logic of stable truth is. The repercussions of this are considerable.

By way of further unwrapping the content of the theory **M** we consider the axiom system **TP**. We again define $P(t) =_{def} T(t) \vee F(t)$, i.e. propositions are those objects which are true or false. Under the present interpretation propositions are those objects which are stable. The first result parallels that for **KFG**.

THEOREM 4.7
The axiom system **TP** is derivable in **M**.

PROOF For propositions we illustrate with rule (v) of Theorem 3.9. The assumption yields $\forall x(T(Ax) \vee T(\sim Ax))$. This implies $\forall x T(Ax) \vee \exists x T(\sim Ax)$. By BAR we obtain $T(\forall xA) \vee \exists x T(\sim Ax)$. By the classical truths $\sim At \to \exists x \sim A$, and $\exists x \sim A \to \sim \forall xA$, NEC and IP, we obtain $T(\forall xA) \vee T(\sim \forall xA)$, as required. For the axioms of T we consider the case of negation. $T(\sim A) \to \sim T(A)$ follows from the axiom DIS whereas $\sim T(A) \to T(\sim A)$ follows only because $P(A)$. ●

In parallel to the system arising from **KFG** consider the natural deduction system given on page 53.

Again we assume the normal side conditions on the quantifier rules for existential elimination and universal introduction. These rules provide us with a natural deduction version of a logic of truth. The rules are not carbon copies of the rules for the classical Predicate Calculus because of the additional conditions regarding stability/propositions in the negation rules, the implication introduction rule, and the disjunction and existential elimination rules. Observe the difference with the rules derived from the system **KFG**; in the latter there are no side conditions on propositions for the disjunction and existential elimination rules. In this respect the above system is more uniform.

THEOREM 4.8
The above proof system is derivable in **M**.

PROOF More exactly we mean that if the premises are derivable then the conclusions are. We argue from **M**. The details are tedious but simple to check. We illustrate with the rule for implication introduction. Given $P(A)$ there are two possibilities: $T(A)$ or $T(\sim A)$. Assume $T(A)$. Use the tautology $B \to (A \to B)$ and NEC to derive $T(B \to (A \to B))$. Under the assumption $T(A)$ we can derive from assumptions of the rule that $T(B)$. Now employ IP to derive $T(A \to B)$. On the other hand, if $T(\sim A)$ then employ the tautology $\sim A \to (A \to B)$, NEC and IP to get $T(A \to B)$. ●

Comparison between this system and the corresponding system for **KFG** should not lead the reader to believe that **M** is weaker than **KFG**. They are orthogonal. These proof systems are not equivalent to the corresponding logics of truth; the proof systems are merely derivable from them.

4.4   Stable Axioms and Necessitation

This has the beginnings of a quite interesting logic of truth but from a modal perspective the above theory is a very weak one. What happens when we try to strengthen it? There are two obvious ways of achieving this: one corresponds to the addition of further modal axioms and the other to the rule of necessitation. We consider the latter first.

The modal logic which results from permitting the full rule of necessitation is the quantifier version of the classical deontic logic **D** (plus Barcan). This is defined by the following axioms and rules:

THE LOGIC **D**

DIS $T(A) \rightarrow C(A)$
IP $T(A \rightarrow B) \rightarrow (T(A) \rightarrow T(B))$
BAR $\forall x T(A) \rightarrow T(\forall x A)$
NEC   If $D \vdash A$ then $D \vdash T(A)$

To establish that each of the theorems of **D** are safe, we first establish that each of the axioms DIS, IP and BAR is not just safe (e.g. DIS is true at stabilization ordinals) but stably true (e.g. T(DIS) is true at stabilization ordinals). We thus need to establish the safeness of the following axioms:

SDIS $T(T(A) \rightarrow C(A))$
SIP $T(T(A \rightarrow B) \rightarrow (T(A) \rightarrow T(B)))$
SBAR $T(\forall x T(A) \rightarrow T(\forall x A))$

THEOREM 4.9
SIP, SDIS and SBAR are safe. (i.e. IP, DIS and BAR are stably true).

PROOF   For SIP we have to show that $T(T(A \rightarrow B) \rightarrow (T(A) \rightarrow T(B)))$ is true at any stabilization ordinal. First note that $T(A \rightarrow B) \rightarrow (T(A) \rightarrow T(B))$ is true at any successor ordinal, by the definition of revision. Moreover at limit ordinals if $A \rightarrow B$ has been true from some ordinal less than the limit ordinal, and A likewise, then B must have been true from the greater of the two ordinals, and thus is true at the limit. For the stability of DIS we have only to observe that

Conjunction
$$\frac{T(A) \qquad T(B)}{T(A \& B)} \qquad \frac{T(A \& B)}{T(A)} \qquad \frac{T(A \& B)}{T(B)}$$

Implication
$$\begin{array}{c} [T(A)] \\ \vdots \\ \frac{T(B) \qquad P(A)}{T(A \to B)} \end{array} \qquad \frac{T(A) \qquad T(A \to B)}{T(B)}$$

Disjunction
$$\frac{T(A)}{T(A \vee B)} \qquad \qquad \begin{array}{cc} & [T(A)] \quad [T(B)] \\ & \vdots \qquad \vdots \\ \frac{P(A) \qquad T(A \vee B) \qquad T(C) \qquad T(C)}{T(C)} \end{array}$$

$$\frac{T(B)}{T(A \vee B)} \qquad \qquad \begin{array}{cc} & [T(A)] \quad [T(B)] \\ & \vdots \qquad \vdots \\ \frac{P(B) \qquad T(A \vee B) \qquad T(C) \qquad T(C)}{T(C)} \end{array}$$

Negation
$$\begin{array}{c} [T(A)] \\ \vdots \\ \frac{\bot \qquad P(A)}{T(\sim A)} \end{array} \qquad \frac{T(A) \qquad T(\sim A)}{\bot}$$

Classical Negation
$$\begin{array}{c} [T(\sim A)] \\ \vdots \\ \frac{\bot \qquad P(A)}{T(A)} \end{array}$$

Universal Quantification
$$\frac{T(A[c])}{T(\forall xA)} \qquad \frac{T(\forall xA)}{T(A[t])}$$

Existential Quantification
$$\frac{T(A[t])}{T(\exists xA)} \qquad \qquad \begin{array}{c} [T(A[c])] \\ \vdots \\ \frac{\forall xP(A) \qquad T(\exists xA) \qquad T(C)}{T(C)} \end{array}$$

T(A) always excludes the possibility of T(~A) at both successor ordinals and limits. The argument for Barcan again relies on the constancy of the domain of individuals. ●

As a direct consequence we have the stability of **D**.

THEOREM 4.10
If **D** ⊢ A then A is stably true.

PROOF   We establish the result by induction on the proofs in **D**. First observe that all the proof rules of the classical Predicate Calculus preserve stability. If A is an instance of any of the axioms of **D** then the result follows from the previous theorem. Finally consider NEC. Here we only have to observe that if a wff is stably true then it is stably true that it is stably true. This follows essentially from the definition of revision at successor ordinals. ●

As a corollary this result also informs us that every theorem of **M** is actually stably true.

COROLLARY 4.11
If **M** ⊢ A then A is stably true.

In the theory **D** we have a full principle of substitution:

   Sub   A ↔ B → Ψ[A] ↔ Ψ[B]

where Ψ is any context in which a wff can be meaningfully substituted. This is easy to prove by induction on the context. The full rule of necessitation plays the crucial role where the context is T itself. We point this out only to compare it with those extensions where only weaker principles will apply.

The main result of this section states that the modal logic **D** is a consistent logic of truth and moreover all the theorems of **D** are stably true. The logic **D** is thus a logic of stable truth. Theorem 4.10 is illustrative of what will become an important distinction. We can prove either that all the theorems of a logic are safe (**M**) or that all the theorems are stably true (**D**). The latter is a stronger result: the former insists only that all the theorems are true at stabilization ordinals whereas the latter demands that they are stably true. In the case of **D** we have no choice since the stronger result is required in order to justify the full rule of necessitation. In general, to establish full necessitation we are forced to prove that all the theorems of the logic are stably true.

From now on we shall employ the terminology that a logic is *safe* if all its theorems are, and a logic is *stably true* if all its theorems are.

4.5   The T-Axiom

The logic **D** is perhaps the simplest useful system. The logic **M**, although interesting, is hard to work with since one has to keep track of which axioms have been used in order to determine whether or not it is permissible to employ necessitation. Unfortunately, not all *natural* axioms will be stably true; some will only be safe. In what follows we shall use classical modal logic as our guide for the search for new axioms. We shall consider the characteristic axioms of the three standard systems:

> **R** $T(A) \rightarrow A$
> **Tr** $T(A) \rightarrow T(T(A))$
> **E** $C(A) \rightarrow T(C(A))$

We employ R, Tr and E for these axioms to keep them separate from the modal logics themselves. We consider these various additional modal axioms beginning with the modal axiom for the modal logic **T**. Here there are some real surprises in comparison with standard modal systems.

Our first result informs us that R is safe.

THEOREM 4.12
R is safe.

PROOF   At stabilization ordinals $T(A)$ states that A is stably true. Observe that if A is stably true then A is true at a stabilization ordinal. This follows because of the rule of revision at successor ordinals and the fact that, if $T(A)$ is true at a stabilization ordinal, then $T(A)$ will be true at its successor.   ●

Of course, R implies DIS but, whereas DIS is stably true, R is not. Indeed, if it were then the modal logic **T** would be a consistent logic of truth, but it is not.

THEOREM 4.13
The modal logic **T** is inconsistent as a logic of truth.

PROOF   From the diagonalization lemma and necessitation there is a wff A such that

> (a) $T(A \leftrightarrow T(\sim A))$

From the tautology $(A \rightarrow \sim A) \rightarrow \sim A$ and necessitation we obtain

> (b) $T((A \rightarrow \sim A) \rightarrow \sim A)$

From the tautology

$$[(A \rightarrow \sim A) \rightarrow \sim A] \rightarrow [(A \leftrightarrow T(\sim A)) \rightarrow ((T(\sim A) \rightarrow \sim A) \rightarrow \sim A)]$$

necessitation yields

(c) $T([[(A \rightarrow \sim A) \rightarrow \sim A] \rightarrow [(A \leftrightarrow T(\sim A)) \rightarrow ((T(\sim A) \rightarrow \sim A) \rightarrow \sim A)]])$

(a), (b) and (c) yield, by IP,

(d) $T((T(\sim A) \rightarrow \sim A) \rightarrow \sim A)$

Employing the axiom R, necessitation and IP, from (d) we obtain

(e) $T(\sim A)$

By (e) and the axiom R we obtain

(f) $\sim A$

On the other hand, by (a) and the axiom R we obtain

(g) $A \leftrightarrow T(\sim A)$

which by (e) yields

(h) A

(f) and (h) are contradictory. ●

As we shall see in chapter 8, logics even weaker than **T** are inconsistent. This result is essentially the truth-theoretic version of the result of Montague and Kaplan [1960]. Their derivation is based upon the *Hangman* or *Knower* paradox. We shall return to their arguments in chapter 8 when we consider similar logics for the operators of predicative modal logic.

So although the T-axiom, R, is safe it is not stable and so cannot be included in the rule of necessitation: we can add R as an axiom but it must not enter into any instance of the rule of necesitation. This is unfortunate but not that unexpected. Indeed, the explanation for this within the logic of stable truth is illuminating since the stable truth of R actually guarantees that everything is a proposition. This is the content of the following axiom:

S $T(T(A) \rightarrow A) \rightarrow P(A)$

THEOREM 4.14
S is stably true.

PROOF We show S is true from some ordinal onwards.
If α is a successor ordinal, $\beta + 1$, then $A \vee \sim A$ will be true at β and so

T(A) ∨ T(~A) (i.e. P(A)) will be true at α. If α is a limit ordinal and T(T(A) → A) is true at α then T(A) → A will be true from some ordinal β(< α) to α. If A is true anywhere between β and α then T(A) will be true at its successor and so A will be true at its successor. So if A is true anywhere between β and α it is true everywhere. Since A is true or ~A is true for all these ordinals it follows that T(A) is true at α or T(~A) is true at α, i.e. P(A) is true at α. Hence, T(T(A) → A) → P(A) is true at every ordinal after the first. Consequently it is stably true. ●

Axiom S blocks the possibility of R being stably true since the stability of R would render everything a proposition. Indeed, this is the underlying semantic explanation of the inconsistency of the logic T. Despite this, however, something quite elegant can be salvaged from the ashes of T-modal logic. Consider the following logic of truth.

THE LOGIC ST

DIS T(A) → C(A)
IP   T(A → B) → (T(A) → T(B))
BAR ∀xT(A) → T(∀xA)
S   T(T(A) → A) → P(A)
NEC If ST ⊢ A then ST ⊢ T(A)

As a consequence of the previous theorem we have:

THEOREM 4.15
If ST ⊢ A then A is stably true.

This is a fascinating result. We cannot have the modal logic T since the characteristic axiom offends — it is safe but generally unstable. Moreover, any instance of it which is found to be stable renders the wff of the instance a proposition. What makes this result interesting concerns its obvious similarity to the provability logic G (Boolos[1979] and Smorynski[1985]) where S is replaced by

T(T(A) → A) → T(A)

These connections are obviously of some interest. Perhaps the most fruitful line of research involves the development of a possible world semantics for D and ST. We shall explore such issues on another occasion. Instead we turn our attention to the modal logic S4. As we shall see, an identical pattern emerges.

4.6   The S4-Axiom

The characteristic axiom of **S4** modal logic takes the form:

Tr $T(A) \to T(T(A))$

As with the characteristic axiom of **T** modal logic, this axiom is safe.

THEOREM 4.16
Tr is safe.

PROOF   Tr insists that if A is stably true then it is stably true that it is stably true. To see that this is so one only has to observe that by the rule of revision at successor ordinals the positive stability of A implies the positive stability of $T(A)$.   ●

R and Tr enable the derivation of $T(A) \leftrightarrow T(T(A))$ which renders iterated applications of the truth predicate redundant.

Once again Tr, although safe, is not stably true. Indeed, we have a result which parallels the T-case. Consider the axiom:

Q $T[T(A) \to T(T(A))] \to P(A)$

This states that the stability of the S4-axiom would render everything a proposition. In parallel to the T-case we have:

THEOREM 4.17
Q is stably true.

PROOF   It is sufficient to show that $T[T(A) \to T(T(A))] \to P(A)$ is true from some ordinal. It is clear from the definition of revision at successor ordinals that $T[T(A) \to T(T(A))] \to T(A \to T(A))$ is true from some ordinal. This reduces the problem to showing that $T(A \to T(A)) \to P(A)$ is true from some ordinal. At successor ordinals the result is immediate from the definition of revision. If λ is a limit ordinal and $T(A \to T(A))$ is true at λ, then there exists an ordinal $\alpha < \lambda$ such that for all β, $\alpha \leqslant \beta < \lambda$, $A \to T(A)$ is true at β. Suppose for some such β, $\sim A$ is true at β. Then $\sim A$ is true at $\beta + 1$ — otherwise A will be true at $\beta + 1$ and consequently $T(A)$ will be and, by definition of revision, A will be true at β. This contradicts the assumption. It follows that, if $\sim A$ is true at any β between α and λ, it will be true at every such ordinal. Hence, either $\sim A$ is true everywhere between α and λ, or A will be true everywhere between α and λ. It follows that either $T(A)$ is true at λ, or $T(\sim A)$ is true at λ.   ●

Once again the S4-axiom can be added to the logic but it must not enter into any proofs involving necessitation. However, just as before we can postulate a stable logic with full necessitation involving both S and Q.

THE LOGIC SS4

DIS	$T(A) \rightarrow C(A)$
IP	$T(A \rightarrow B) \rightarrow (T(A) \rightarrow T(B))$
BAR	$\forall x T(A) \rightarrow T(\forall x A)$
S	$T(T(A) \rightarrow A) \rightarrow P(A)$
Q	$T[T(A) \rightarrow T(T(A))] \rightarrow P(A)$
NEC	If $\text{SS4} \vdash A$ then $\text{SS4} \vdash T(A)$

It follows from the previous theorem that this logic is stable.

THEOREM 4.18
If $\text{SS4} \vdash A$ then A is stably true.

So, although we cannot have the full logics **T** and **S4**, we can have these variations where the stability of the corresponding axioms imply propositionhood for the sentence in question. It appears that a natural pattern is emerging, but **S5** provides a slight variation.

4.7 The S5-Axiom

The analogous S5-axiom is not even safe. In the present context it takes the form:

$$\text{E}\quad C(A) \rightarrow T(C(A))$$

In terms of stability, E is clearly false since it insists that any assertion which is not stably false, is stably not stably false. Indeed, we have:

THEOREM 4.19
$\text{ST} + \text{E}$ is inconsistent.

PROOF From E and R we obtain $\sim T(\sim A) \leftrightarrow T(\sim T(\sim A))$. Unfortunately, the *Liar* sentence satisfies $T(B \leftrightarrow T(\sim B))$, and so we obtain, writing B for A in the above, the following equivalences:

$$\sim T(\sim B) \leftrightarrow T(\sim T(\sim B)) \leftrightarrow T(\sim B) \quad \bullet$$

The analogy between E and the T- and S4-axioms is not exact. However, the fact that the S5-axiom is not safe suggests that we examine the following:

W $[C(A) \rightarrow T(C(A))] \rightarrow P(A)$

Not too surprisingly, we have:

THEOREM 4.20
W is stably true.

PROOF We show that W is true from some ordinal onwards. It is clearly true at all successor ordinals. Assume that $C(A) \rightarrow T(C(A))$ is true at a limit ordinal λ. If $C(A)$ is true at λ then $T(C(A))$ will be, hence there is an ordinal $\alpha < \lambda$ such that $C(A)$ is true between α and λ. But now $\sim A$ cannot be true at any such intermediate ordinal and so A must be true for all such intermediate ordinals. Hence $T(A)$ is true at λ. ●

This brings us to our analogy for **S5** modal logic.

THE LOGIC SS5

DIS	$T(A) \rightarrow C(A)$
IP	$T(A \rightarrow B) \rightarrow (T(A) \rightarrow T(B))$
BAR	$\forall x T(A) \rightarrow T(\forall x A)$
S	$T(T(A) \rightarrow A) \rightarrow P(A)$
Q	$T[T(A) \rightarrow T(T(A))] \rightarrow P(A)$
W	$[C(A) \rightarrow T(C(A))] \rightarrow P(A)$
NEC	If $\mathbf{SS5} \vdash A$ then $\mathbf{SS5} \vdash T(A)$

THEOREM 4.21
If $\mathbf{SS5} \vdash A$ then A is stably true.

So the logic analogous to **S5** behaves in a slightly different way to the other logics. There is, however, a general pattern to the differences between the logics of truth and the standard modal systems which is borne out by the correspondence between R, Tr, E on the one hand and S, Q, W on the other. The logics **ST**, **SS4** and **SS5** are, however, formally different to the systems **T**, **S4** and **S5** and really deserve a full investigation. We shall not pause to look more deeply into their fine structure but, instead, investigate a further principle.

4.8 A Further Non-standard Principle

That the modal logic of truth is different to the standard systems of modal logic should now be quite evident. All the axioms we have considered for inclusion in our logics of truth have been valid in some standard system of modal logic but the modifications we made to T, S4 and S5 were highly non-standard.

We now consider a further non-standard axiom:

IPT $T(T(A) \rightarrow T(B)) \rightarrow T(A \rightarrow B)$

This is certainly not valid in any standard modal logic. It is nevertheless safe.

THEOREM 4.22
IPT is safe.

PROOF This is clear from the definition of truth at successor ordinals. ●

On the other hand IPT is not stably true: if it were then applying IPT to the assertion of its stable truth would yield $T((T(A) \rightarrow T(B)) \rightarrow (A \rightarrow B))$ which by R gives $(T(A) \rightarrow T(B)) \rightarrow (A \rightarrow B)$. Any two sentences A and B, where A is true but not stably true and B is false, provide a counterexample to this principle.

This principle does have some unexpected consequences. A special case of IPT is the axiom (with negation interpreted as $A \rightarrow \perp$):

L $T(C(A)) \rightarrow T(A)$

the converse of which, namely:

N $T(A) \rightarrow T(C(A))$

is already derivable in SS4. We shall often refer to the conjunction of N and L as NEG, since they permit negations to be moved through iterated applications of the truth predicate, i.e.

$T(\sim A) \leftrightarrow T(\sim T(A))$

These principles go beyond the standard systems of modal logic but they do not run counter to our intuitions about truth since the "ultimate" logic of truth is that given by the Tarski biconditionals and these justify all modal principles since the logic of modality collapses under them.

4.9 Some Consequences

To complete our discussion we investigate some of the obvious consequences of these logics. For convenience, let **LS** be the logic **SS5** + R + Tr + IPT. Observe that **LS** does not support full necessitation.

THEOREM 4.24

In **LS** we have:

(i) $P(A) \leftrightarrow P(T(A))$
(ii) $P(A) \leftrightarrow T(P(A))$
(iii) $P(A) \leftrightarrow (T(A) \leftrightarrow A \ \& \ F(A) \leftrightarrow \sim A)$
(iv) $P(A) \leftrightarrow T(T(A) \rightarrow A)$

PROOF   For part (i) we use R, Tr and NEG. For the second part we employ R and $T(X) \rightarrow T(X \lor Y)$. Part (iii) employs M and R. For (iv) observe that from $T(A)$, the tautology $A \rightarrow (T(A) \rightarrow A)$, NEC and IP, we obtain $T(T(A) \rightarrow A)$; and from $T(\sim A)$, NEG, the tautology $\sim B \rightarrow (B \rightarrow C)$, NEC and IP, we again obtain $T(T(A) \rightarrow A)$, S giving the other half.   ●

From (iv) and (iii) we see that $T(T(A) \rightarrow A)$ is both necessary and sufficient for stability and that the positive and negative statements of the Tarski biconditionals are also necessary and sufficient.

The lack of a full axiom of necessitation means that substitution fails in **LS**. However, a weaker principle is derivable. First define

$$A \simeq B =_{\text{def}} T(A \leftrightarrow B)$$

then the following principle:

$$\simeq \text{**Sub**} \quad A \simeq B \rightarrow \Psi[A] \simeq \Psi[B]$$

is derivable, where Ψ is any context in which a wff can be meaningfully substituted.

THEOREM 4.25

The above principle is derivable in **LS**.

PROOF   Use induction on the context. The case where the context is $T(\ )$ is taken care of by SIP.   ●

4.10   The *Liar* and the *Truth Teller*

To complete this chapter we briefly examine the *liar* and the *truth teller* under the government of stable truth. This will bring out their very different behaviour. If you recall, the two sentences have the following content:

| *Liar* | $A \leftrightarrow T(\sim A)$ |
| *Truth Teller* | $B \leftrightarrow T(B)$ |

The *liar* says of itself that it is false whereas the *truth teller* claims of itself that it is true. We first examine their behaviour within the model-theoretic setting of the revision process.

We have already indicated that the *liar* oscillates in truth value: if A is initially true then at the first revision it switches truth value and is false; at the second revision it is true again and so on. The *truth teller*, on the other hand, maintains its truth value. If B is initially true then T(B) will be true at the first revision and, by T(B) $\leftrightarrow$ B, B will be true at this model and so on. If B is false at the initial model then it will remain so throughout the revision process. Consequently, B is stable.

We next examine how this semantic account is reflected within the logic of stable truth.

THEOREM 4.26
In the logic of stable truth the *liar* is paradoxical and the *truth teller* is a proposition.

PROOF   We tackle the *liar* first. If A is a proposition then $T(A) \vee T(\sim A)$. If the first disjunct holds then, by R, A follows. By $A \leftrightarrow T(\sim A)$ we obtain $T(\sim A)$. By DIS we obtain a contradiction. If the second disjunct holds, i.e. $T(\sim A)$, then, by R, $\sim A$, but this contradicts $T(\sim A) \leftrightarrow A$. Hence $\sim P(A)$, i.e. A is paradoxical.

Next consider the *truth teller*. This is constructed by diagonalization as $A = T(t)$ where $t = T(t)$. By necessitation $T(t = T(t))$. Next observe that from the equality rules of Predicate Calculus and necessitation:

$$T(a = b) \rightarrow [T(a) \leftrightarrow T(b)]$$

Hence, $T(T(t) \leftrightarrow T(T(t)))$, i.e. $T(A \leftrightarrow T(A))$. By the S-axiom we obtain P(A).

●

This result has more general implications. It establishes that self-referential sentences are not all equally problematic and some are even propositions.

The logic of stable truth is worthy of further attention. We have left many questions unanswered and in particular any detailed possible world semantic analysis of the various modal (ST, SS4, SS5) logics and their connections with provability logics. We trust we have provided the reader with material for further investigation.

5 Frege Structures

The final theory of truth that we shall consider is due to Scott[1975] and Aczel[1980]. This theory is foreshadowed by the theory of truth and propositions we called **TP**. In the previous chapters we established that both the fixpoint and stable theories of truth support the theory **TP**. Moreover, **TP** seems the least we might demand of a theory of truth and propositions: the axioms of propositions insist that the notion of proposition is preserved by the logical connectives and quantifiers and the axioms of truth simply unwind the Tarski truth conditions on propositions. As we shall see, **TP** forms the backbone of the Scott–Aczel theory. We first present the version of the Scott–Aczel theory which is due to Aczel and then connect it with our theory **TP**.

5.1 The Language of Frege Structures

Actually, the theory of Aczel[1980] is only a slight variation on **TP**. The major difference concerns the setting of the theory: it is cast within the language of combinatory logic. The pure Lambda Calculus is enriched by the addition of *logical combinators* which are intuitively the *intensional* analogues of the logical connectives and quantifiers. Formally, the language of the theory, $(L2)$, has the following structure.

BASIC VOCABULARY OF $L2$
Individual variables $x, y, z, \ldots$
Individual constants   c, d, e, ...
Logical combinators $\wedge, \vee, \daleth, \Rightarrow, \Leftrightarrow, \varXi, \Theta, \approx$

INDUCTIVE DEFINITION OF TERMS
    (i)   Every variable, constant or logical combinator, is a term.
    (ii)   If t is a term and x is a variable then $(\lambda x . t)$ is a term.
    (iii) If t and t' are terms then (tt') is a term.

INDUCTIVE DEFINITION OF WELL-FORMED FORMULAE

    (i)   If t and s are terms then $s = t$, $\text{Prop}(t)$ and $\text{True}(t)$ are (atomic) wff.

    (ii)   If Φ and Φ' are wff then $\Phi \& \Phi'$, $\Phi \lor \Phi'$, $\Phi \to \Phi'$, $\sim \Phi$ are wff.

    (iii)   If Φ is a wff and x a variable then $\exists x \Phi$ and $\forall x \Phi$ are wff.

To the reader unfamiliar with combinatory logic this language may appear somewhat strange. The language of terms contains, in addition to the constructs of the pure Lambda Calculus, the logical combinators. According to one class of intuitions these are the *intensional* analogues of the connectives and quantifiers: $\land$, $\lor$, $\daleth$, $\Rightarrow$, $\approx$ are respectively the intensional connectives of conjunction, disjunction, negation, implication, and intensional identity; Ξ and Θ are, respectively, the intensional existential and universal quantifiers. We shall treat $\Leftrightarrow$ as defined in terms of implication in the obvious way. The language of terms thus permits the intensional representation of logical assertions through these intensional connectives and quantifiers. For example, $\Theta(\lambda x.(\approx xx))$ is a well-formed term which is the *intensional* analogue of the assertion that every object is equal to itself. We shall often write $\Rightarrow xx$, $\lor xy$ as $x \Rightarrow x$ and $x \lor y$, etc. These logical combinators facilitate the coding of wff as terms which was present in our language L. The reference to *intensionality* might be taken simply as a means of distinguishing between these shadowy representations of wff and the actual wff themselves, given by the language of wff. The latter are *extensional* in that they are to be directly interpreted as assertions which are either true or false. The *intensional* assertions, on the other hand, only obtain their truth values through application of the truth predicate. Moreover, not all terms can be meaningfully taken to possess a truth value. Many terms will make no syntactic sense when interpereted as logical assertions. Indeed, even some terms which pass the grammatical test will turn out to be problematic for paradoxical reasons. The predicate Prop is there to pick out the legitimate terms which can be safely taken to be propositions. In one sense the language of wff acts as the metalanguage for the language of terms. In another, they form an *intensional/extensional* pair. The term *intensional* is not inappropriate here since the equality relation on terms is that of the Lambda Calculus and two terms which can be taken to be propositions will only be equal when this is derivable within the Lambda Calculus. This is a highly intensional notion of equality. In particular,

    (1) $T(t) \leftrightarrow T(s)$

does not imply

    (2) $t = s$

66

Having the same truth-value is not sufficient to yield the conclusion that the terms are equal.

The reader should now be able to tease out the differences between $L2$ and L: in the latter, the connectives and quantifiers occur as terms *only* in the context of (coded) wff. This completes the syntax and brings us to the axiomatic theory itself.

The actual theory pertains to the logic of the predicates Prop and True. The former asserts that something is a proposition and the latter asserts that it is a true proposition. In the Aczel theory these predicates are governed by the following axioms.

AXIOMS OF PROPOSITIONS
(i) $(\text{Prop}(t) \& \text{Prop}(s)) \rightarrow \text{Prop}(t \wedge s)$
(ii) $(\text{Prop}(t) \& \text{Prop}(s)) \rightarrow \text{Prop}(t \vee s)$
(iii) $(\text{Prop}(t) \& (\text{True}(t) \rightarrow \text{Prop}(s))) \rightarrow \text{Prop}(t \Rightarrow s)$
(iv) $\text{Prop}(t) \rightarrow \text{Prop}(\daleth t)$
(v) $\forall x \text{Prop}(t) \rightarrow \text{Prop}(\Theta(\lambda x.t))$
(vi) $\forall x \text{Prop}(t) \rightarrow \text{Prop}(\Xi(\lambda x.t))$
(vii) $\text{Prop}(s \approx t)$

Most of these are self-explanatory and parallel the axioms of **TP**. Together they simply state that the notion of proposition is closed under the intensional connectives and quantifiers. We shall often abbreviate $\Xi(\lambda x.t)$ as Ξxt, etc. The last axiom insists that the assertion of intensional equality constitutes a proposition. The axioms of truth flow naturally from the closure conditions on propositions.

AXIOMS OF TRUTH
(i) $\text{Prop}(t) \& \text{Prop}(s) \rightarrow (\text{True}(t \wedge x) \leftrightarrow \text{True}(t) \& \text{True}(s))$
(ii) $\text{Prop}(t) \& \text{Prop}(s) \rightarrow (\text{True}(t \vee s) \leftrightarrow \text{True}(t) \vee \text{True}(s))$
(iii) $\text{Prop}(t) \& (\text{True}(t) \rightarrow \text{Prop}(s)) \rightarrow (\text{True}(t \Rightarrow s) \leftrightarrow (\text{True}(t) \rightarrow \text{True}(s)))$
(iv) $\text{Prop}(t) \rightarrow (\text{True}(\daleth t) \leftrightarrow \sim \text{True}(t))$
(v) $\forall x \text{Prop}(t) \rightarrow (\text{True}(\Theta(\lambda x.t)) \leftrightarrow \forall x \text{True}(t))$
(vi) $\forall x \text{Prop}(t) \rightarrow (\text{True}(\Xi(\lambda x.t)) \leftrightarrow \exists x \text{True}(t))$
(vii) $\text{True}(s \approx t) \leftrightarrow s = t$
(viii) $\text{True}(t) \rightarrow \text{Prop}(t)$

To restate the obvious: the axioms for truth apply only to those objects which are propositions but on the class of propositions they yield the standard Tarski truth conditions. The axioms thus provide us with a formal version of the notion of proposition and its standard Tarski truth conditions. The last axiom informs us that everything which is true is a proposition. This theory is natural and easy to grasp and we therefore hope that little more needs to be said by way of explanation. We are, however, under some obligation to spell out its connection with the theory **TP** and hence with the two major theories of the previous chapters.

5.2   Models in **TP**

That the theory of Frege Structures can be modelled in **TP** should come as no surprise since the axioms of **TP** are parallel to those of a Frege Structure. Nevertheless, for completeness we spell out the details.

The main technical consideration relates to how the logical combinators of a Frege Structure are to be represented in the language L.

DEFINITION 5.0
Let t and s be terms of L. Then define

(i)	$t \wedge s$	$= \hat{\ }(T(t) \& T(s))$
(ii)	$t \vee s$	$= \hat{\ }(T(t) \vee T(s))$
(iii)	$t \Rightarrow s$	$= \hat{\ }(T(t) \rightarrow T(s))$
(iv)	$\daleth t$	$= \hat{\ }(\sim T(t))$
(v)	Ξf	$= \hat{\ }(\exists x T(fx))$
(vi)	Θf	$= \hat{\ }(\forall x T(fx))$
(vii)	$t \approx s$	$= \hat{\ }(t = s)$

The combinators are thus introduced via the truth predicate and the formation of terms from wff. In addition, we must define the predicates True and Prop. These definitions are also quite straightforward and capture the ideas of groundedness/stability when interpreted in the appropriate theory.

DEFINITION 5.1

(i)	True(t)	$= T(T(t))$
(ii)	False(t)	$= F(T(t))$
(iii)	Prop(t)	$= \text{True}(t) \vee \text{False}(t)$

In the theory **KFG**, True(t) says that $T(t)$ is positively grounded and False(t) that $T(t)$ is negatively grounded; in the logic of stable truth, groundedness is replaced by stability. The iterated T's in (i) may appear somewhat odd. The reason is that the axioms of truth, from the previous theories, apply primarily to wff not terms. We wish to interpret True(t) as t is grounded/stable but since these notions are primarily notions which pertain to wff we have to iterate the truth predicate.

The following is an immediate consequence of 5.1.

THEOREM 5.2

Under the definitions 5.0 and 5.1 the axioms of a Frege Structure **FS** follow from the logic of truth **TP**.

PROOF   This follows immediately from definitions and the axioms of propositions and truth which constitute **TP**.   ●

This section, although technically quite straightforward, indicates how the three theories are hooked together and how the logic of truth which underpins Frege Structures is supported by both the fixpoint and stable theories of truth.

5.3   Properties and Relations

The minimal theory **TP**, within its present guise as a Frege Structure, constitutes a rather elegant theory of *truth* and *propositions*. In addition, it supports an interesting theory of *properties* and *relations*. To this we now turn.

The main theme of *property theory* concerns the development of the notions of *proposition, property* and *relation*. In the setting of Frege Structures the first notion is taken as primitive while the other two are derived. Essentially, properties and relations are introduced as propositional functions where the notion of function is inherited from that of the Lambda Calculus.

DEFINITION 5.3

$$\text{Rel}_1(f) =_{\text{def}} \forall x(\text{Prop}(fx))$$
$$\text{Rel}_{n+1}(f) =_{\text{def}} \forall x(\text{Rel}_n(fx))$$

Relations are defined in curried form by recursion. We shall write Pty (*property*) for Rel_1. We shall often write $\lambda x.t$ as $\{x:t\}$ and $\text{True}(tx)$ as $x\varepsilon t$, especially when t is a property.

Properties and relations like propositions are taken to be intensional notions. Two properties will not be the same just because they apply to exactly the same objects. Properties unlike sets are not characterized extensionally in terms of the objects which they delineate. In other words, for properties the axiom of extensionality:

$$\forall z(z\varepsilon\{x:t\} \leftrightarrow z\varepsilon\{x:s\}) \rightarrow s = t$$

fails. This is because equality is that given by the Lambda Calculus and it is perfectly possible for two Lambda abstracts to yield the same truth values for all arguments without being equal as terms. In terms of the intended application of these theories to knowledge representation and natural language semantics these are highly desirable qualities.

THEOREM 5.4

    (i) $\forall z(z\varepsilon\{x:t\} \leftrightarrow \text{True}(t[z/x]))$

    (ii) $\text{Pty}(\lambda x.t) \rightarrow$

        $[\text{True}(\Theta z((\lambda x.t)z \Leftrightarrow t[z/x])) \leftrightarrow \forall z(z\varepsilon\{x:t\} \leftrightarrow \text{True}(t[z/x]))]$

PROOF   (i) follows from the equality rule of the predicate Calculus. (ii) follows directly from the assumption and the axioms of propositions and truth: $\text{Pty}(\lambda x.t)$ implies $\text{Prop}((\lambda x.t)z)$ and $\text{Prop}(t[z/x])$, for all z. Hence, for all z, $\text{Prop}((\lambda x.t)z \Leftrightarrow t[z/x])$ and so

    $\text{Prop}(\Theta z((\lambda x.t)z \Leftrightarrow t[z/x]))$

The rest now follows from the axioms of truth.   ●

Part (i) is an axiom of comprehension for properties and sanctions the expected equivalence between set-membership and substitution and (ii) informs us that, for abstracts which are properties, the intensional equivalence

    $\Theta z((\lambda x.t)z \Leftrightarrow t[z/x])$

has the correct truth conditions.

We next turn to the development of the implicit theory of properties and relations. We first examine some fairly standard operations on properties. These correspond to the obvious set-theoretic constructions.

DEFINITION 5.5

    (i) $\cap = \lambda f.\lambda g.\{x: fx \wedge gx\}$

    (ii) $\cup = \lambda f.\lambda g.\{x: fx \vee gx\}$

    (iii) $- = \lambda f.\lambda g.\{x: fx \wedge \neg(gx)\}$

    (iv) $\otimes = \lambda f.\lambda g.\{z: \Xi x \Xi y(z \approx \langle x, y\rangle \wedge fx \wedge gy)\}$

    (v) $\oplus = \lambda f.\lambda g.\{z: (\text{fst}(z) \approx 0 \wedge f(\text{snd}(z))) \vee (\text{fst}(z) \approx 1 \wedge g(\text{snd}(z)))\}$

    (vi) $\text{-}\!\gg = \lambda f.\lambda g.\{z: \Theta x(fx \Rightarrow g(zx))\}$

These combinators correspond to union, intersection, complementation, disjoint union, cartesian product, and function space respectively. Moreover, we can derive the following closure conditions for these constructors where we shall write $\cap xy$ as $x \cap y$, etc.

THEOREM 5.6

For O any of the combinators $\cap$, $\cup$, $-$, $\oplus$, $\otimes$, $-\!\!\gg$,

$$\text{Pty}(t) \,\&\, \text{Pty}(s) \to \text{Pty}(tOs)$$

PROOF   All the results follow from the rules for propositions and are straightforward if tedious to check. We illustrate with $\cap$. Suppose $\text{Pty}(f)$ and $\text{Pty}(g)$. Then by definition of Pty we have for each x, $\text{Prop}(fx)$ and $\text{Prop}(gx)$. By the closure axioms for propositions, for each x, $\text{Prop}(fx \wedge gx)$. Hence, $\text{Pty}(\lambda x.(fx \wedge gx))$.   ●

As a further immediate consequence we have the natural truth conditions for these constructs.

THEOREM 5.7

For $\text{Pty}(t)$ and $\text{Pty}(s)$ we have:

(i) $z\varepsilon(t \cap s) \;\leftrightarrow z\varepsilon t \,\&\, z\varepsilon s$
(ii) $z\varepsilon(t \cup s) \;\leftrightarrow z\varepsilon t \lor z\varepsilon s$
(iii) $z\varepsilon(t - s) \;\leftrightarrow z\varepsilon t \,\&\, \sim(z\varepsilon s)$
(iv) $z\varepsilon(t \otimes s) \;\leftrightarrow \text{fst}(z)\varepsilon t \,\&\, \text{snd}(z)\varepsilon s$
(v) $z\varepsilon(t \oplus s) \;\leftrightarrow (\text{fst}(z) = 0 \,\&\, \text{snd}(z)\varepsilon t) \lor (\text{fst}(z) = 1 \,\&\, \text{snd}(z)\varepsilon s)$
(vi) $z\varepsilon(t -\!\!\gg s) \;\leftrightarrow \forall x(x\varepsilon t \to z x\varepsilon s)$

PROOF   Use 5.4 part (i) together with the observation that t and s are properties. We then apply the axioms for truth.   ●

The next combinators provide the universal and empty properties respectively. This is perhaps the most surprising consequence of this theory since the existence of the universal property is not a familiar consequence of most set/class theories.

DEFINITION 5.8

(i) $\nabla = \{x : (x \approx x)\}$ (ii) $\Omega = \{x : \neg(x \approx x)\}$

The following establishes our right to call these properties.

THEOREM 5.9

$$\text{Pty}(\nabla) \,\&\, \text{Pty}(\Omega)$$

PROOF   These follow from the propositional nature of intensional equality.   ●

In fact there are many universal properties since each intensional logical truth will generate one. They are of course all extensionally equivalent.

The definitions and results 5.5–5.9 establish that the theory is a quite useful one in that they establish closure under some of the standard set-forming constructors. Indeed, the theory given by 5.5–5.9 is "elementary" in the sense of Feferman[1979]. However, the theory supports more than these constructions. For our next examples we turn to Martin-Löf[1979] and introduce the notions of dependent product and sum for properties. These also correspond to Feferman's[1979] *Join* axiom.

DEFINITION 5.10

(i) $\Pi = \lambda f \lambda g. \{h : \Theta x (fx \Rightarrow gx(hx))\}$

(ii) $\Sigma = \lambda f \lambda g. \{h : f(\mathrm{fst}(h)) \wedge g(\mathrm{fst}(h))(\mathrm{snd}(h))\}$

THEOREM 5.11
Suppose $\mathrm{Pty}(f)$ and $\forall x(x\varepsilon f \rightarrow \mathrm{Pty}(gx))$. Then

(i) $\mathrm{Pty}(\Pi fg)$ and $h\varepsilon\Pi fg \leftrightarrow \forall x(x\varepsilon f \rightarrow hx\varepsilon gx)$

(ii) $\mathrm{Pty}(\Sigma fg)$ and $h\varepsilon\Sigma fg \leftrightarrow \mathrm{fst}(h)\varepsilon f \,\&\, \mathrm{snd}(h)\varepsilon g(\mathrm{fst}(h))$

PROOF We illustrate with Π. We are given that $\forall x(x\varepsilon f \rightarrow \mathrm{Pty}(gx))$ and $\mathrm{True}(fx)$ implies $\mathrm{Prop}(gx(hx))$. Hence for all x, $\mathrm{Prop}(fx \Rightarrow gx(hx))$. This is what is required since this implies

$\mathrm{Prop}(\Theta x(fx \Rightarrow gx(hx)))$

The truth conditions are then reasonably automatic. ●

These dependent type constructors form the basis of Martin-Löf's theory of types. The closure of the notion of property under these constructions suggests that there is a reasonably rich and useful property/intensional set-theory buried here. We shall not pursue this topic further; we have done enough to give the reader a feeling for what can be obtained. We now briefly review one application of the theory.

5.4 Natural Language Semantics

An application of this material which is worthy of mention concerns its employment in natural language semantics. We have already indicated that the theory is of a highly intensional kind. Neither propositions nor properties are identified via extensional criteria: propositions can share the same truth value without being identical and properties can characterize exactly the same objects and yet be distinct. Much of the recent criticism of the Montague paradigm in formal semantics emanates from the coarse-grained notions of proposition and property which support the theory. As we stated in the introduction, the notion of proposition as a set of possible worlds seems too coarse-grained a notion on which to base a theory of intensionality. The present theory looks like a promising candidate on which to build an intensional theory of semantics (and indeed as a potential language and theory of knowledge representation).

To implement such a proposal is a relatively straightforward affair. The general plan is as follows. Natural language sentences and their constituents correspond, in intension, to the objects of the Frege structure. In particular, sentences are intensionally unpacked as propositions and predicative phrases, such as verbs and adjectives, as properties. All this can be achieved in a straightforward manner simply by shadowing the Montague account. In Montague's system constructs such as determiners, quantifiers, adverbs, prepositions, etc. all get their intensions as higher-type Lambda expressions. In the present theory these are not explicitly available to us in their typed versions. However, they can be represented within the theory in their untyped format. We indicate how this is achieved by developing the theory of quantifiers and determiners.

First we provide the standard definition of the determiners.

DEFINITION 5.12

(i) $Every$ $= \lambda x.\lambda y. [\Theta z(xz \Rightarrow yz)]$
(ii) $Some$ $= \lambda x.\lambda y. [\Xi z(xz \wedge yz)]$
(iii) No $= \lambda x.\lambda y. [\Theta z(xz \Rightarrow \daleth (yz))]$
(iv) $Notall$ $= \lambda x.\lambda y. [\Xi z(xz \wedge \daleth (yz))]$
(v) The $= \lambda x.\lambda y. [\Xi z(xz \wedge yz \wedge \Theta u(xu \Rightarrow u \approx z))]$

These definitions just shadow those of Montague. There are differences but these reflect how the intensional information is coded in Montague's intensional logic and the untyped nature of the present theory.

THEOREM 5.13
For d any of the above determiners we have:

$$Pty(s) \& Pty(t) \to Prop(dst)$$

PROOF   The result follows directly from the axioms of Prop. But we illustrate with (i). $Pty(x)$ and $Pty(y)$ implies that $Prop(xz)$ and $Prop(yz)$ for each z. Hence, for each z,

$\text{Prop}(xz \Rightarrow yz)$ and so $\text{Pty}(\Theta z(xz \Rightarrow yz))$ ●

This is the result we require. It informs us that these determiners, when given properties as arguments, yield propositions as a result.

The quantifiers and determiners also admit of the obvious truth conditions.

THEOREM 5.14

For $\text{Pty}(s)$ and $\text{Pty}(t)$ we have:

(i) $\text{True}(Every\ st) \leftrightarrow \forall x(x\varepsilon s \to x\varepsilon t)$
(ii) $\text{True}(Some\ st) \leftrightarrow \exists x(x\varepsilon s\ \&\ x\varepsilon t)$
(iii) $\text{True}(No\ st)\quad\ \leftrightarrow \forall x(x\varepsilon s \to\ \sim(x\varepsilon t))$
(iv) $\text{True}(Notall\ st) \leftrightarrow \exists x(x\varepsilon s\ \&\sim(x\varepsilon t))$
(v) $\text{True}(The\ st)\quad\ \leftrightarrow \exists x(x\varepsilon s\ \&\ x\varepsilon t\ \&\ \forall u(u\varepsilon s \to u = x))$

PROOF   Use the fact that $\text{Pty}(s)$ and $\text{Pty}(t)$ to ensure that all the atomic assertions are propositions; then use the axioms of True.   ●

Propositions provide the intensions of sentences and their truth-values or extensions are furnished via the truth-predicate (True). Properties provide the intensions of predicative phrases such as verbs and adjectives, and determiners correspond in intension to the determiners of natural language. Indeed, we can introduce the abstract notions of *quantifier* and *determiner* along the same lines as *property* and *relation*.

DEFINITION 5.15

$$\text{Det}(f) =_{\text{def}} \forall x(\text{Pty}(x) \to \text{Quant}(fx))$$
$$\text{Quant}(f) =_{\text{def}} \forall x(\text{Pty}(x) \to \text{Prop}(fx))$$

The notion of quantifier is closed under the Boolean operations introduced for properties. The following is easy to check.

THEOREM 5.16

$$\text{Quant}(f)\ \&\ \text{Quant}(g) \to \text{Quant}(f \cap g)\ \&\ \text{Quant}(f \cup g)\ \&\ \text{Quant}(f - g)$$

Indeed, all of the classical type hierarchy can be reflected in the same fashion. The function space predicates, which reflect the higher type spaces of Montague, can be constructed via the definition:

$$P \equiv\ > Q(f) =_{\text{def}} \forall x(P(x) \to Q(fx))$$

It thus appears that this theory provides a simple elegant and highly intensional first-order theory in which to carry out formal semantics in the Montague style. More details can be found in Turner [1989].

5.5 Undefinability Results

In this final section of this chapter we examine the formal limitations of the theory. These results concern the concept of *internal definability*.

DEFINITION 5.17
A predicate P is ***internally definable*** iff there exists a property p such that

$$\forall x(P(x) \leftrightarrow x \varepsilon p)$$

The final result informs us that the main concepts of the theory are not internally definable.

THEOREM 5.18
The predicates of truth, proposition and property are not internally definable.

    PROOF We illustrate with truth. Assume that the notion of truth is internally definable by the property p.

    Then, (i): $\forall x(\text{True}(\neg(px)) \leftrightarrow \sim\text{True}(px))$, since p is a property. Let $f = \lambda x.\ \neg(px)$ and $t = Yf$. By the fixpoint theorem, let $t = \neg(pt)$. Then by (i), $\text{True}(\neg(pt)) \leftrightarrow \sim\text{True}(pt)$.

    By internal definability of Truth, $\text{True}(x) \leftrightarrow \text{True}(px)$ and, in particular, $\text{True}(t) \leftrightarrow \text{True}(pt)$. Hence $\sim\text{True}(pt) \leftrightarrow \sim\text{True}(t)$.

    Hence, $\sim\text{True}(t) \leftrightarrow \sim\text{True}(pt) \leftrightarrow \text{True}(\neg(pt)) \leftrightarrow \text{True}(t)$, where the last step follows from definition of t. Hence we have a contradiction. ●

This result sets the limits on what can be achieved in the theory of Frege Structures and hence all the theories of truth considered. Intuitively, one might have expected something of the sort to go wrong with the theory but it is unfortunate that the very notions with which the theory is concerned live outside it.

    This completes our discussion of the three theories of truth. We now turn our attention to our second topic, namely modality and its interaction with truth.

6 Modal Logic

In the next three chapters we take up the second theme of the book, namely *modality* and its interaction with *truth*. The literature contains two different approaches to modal logic. The first and most highly developed account views the modal operators as sentential connectives similar to the other logical connectives. Of course, there are major differences between conjunction, disjunction, implication and negation on the one hand and the modal operators on the other, but these differences have to do with the semantics of these connectives, not with the nature of their arguments. In this account both the logical connectives and the modals syntactically select their arguments from the syntactic category of wff. This approach results in traditional modal logic and it is to be seen in contrast to the other, almost abandoned approach, according to which the modal operators are one-place predicates much like the truth predicate itself. This second approach is entirely cast within ordinary predicate logic since the modal operators are predicates rather than new sentential connectives. Modal logic thus becomes a simple first-order theory with axioms which govern the logical behaviour of the modal predicates. This predicative approach has, however, fallen into disrepute since the standard systems of modal logic (T, S4, S5) are inconsistent when modality is recast as a predicate.

The plan for this half of the book is as follows. In this chapter we review the basic background material on traditional modal logic. We present the possible world semantics and the standard proof theoretic systems of classical modal logic. We shall be brief but present all the background necessary to follow the remaining material. Hughes and Cresswell [1968] contains all the details. In the next chapter we develop the three theories of truth within the context of classical modal logic. We add a truth predicate to modal logic and investigate the interaction of the logics of truth and modality. That this might prove to be a fruitful line of research was first proposed in Kripke [1975] where he suggested that his theory of truth might profitably be extended to modal logic. In fact, we will be more ambitious and examine the modal extensions of each of the theories considered. In chapter 8 we turn to the predicative approach to modal logic and use the results and discussion of chapter 7 to guide our search for natural and consistent modal systems of predicative modality.

6.1 The Language

We return to our original language and extend it by the addition of the modal operators. More explicity, the language L is enriched by the addition of the modal operators $\Box$ (it is necessary that) and $\Diamond$ (it is possible that); we add to the language a new clause for wff:

(viii)　If A is a wff then $\Box$A and $\Diamond$A are wff.

The rest of the clauses for L remain intact including the clause for the truth predicate. We shall employ parentheses freely to disambiguate the language. Let $L3$ be the extended language. The coding for the Predicate Calculus is extended to include the modal operators in the obvious way. For those unfamiliar with modal logic it should be stressed that these operators are new sentential connectives not predicates.

6.2 Possible World Semantics

The traditional theory of modal logic is characterized by its employment of the notion of possible world in its semantic foundations. This notion has particular appeal when used to explicate the semantic intuitions underlying the metaphysical modals of necessity and possibility. However, I am less than sanguine about its application to the epistemic and deontic modals. In providing a semantic foundation for the metaphysical modals the notion of alternative possible worlds seems entirely appropriate, and indeed something much like it seems necessary in order to reflect those intuitions which insist that concepts such as "possibility" and "necessity" implicitly appeal to states of affairs which are different to the ones we inhabit. One might go further and advocate that a possible world analysis is also right for a more epistemic interpretation of these modals. What is more controversial is that such an analysis is correct for the deontic modals of belief and knowledge. In these cases the notion of proposition plays the crucial role and its analysis as a set of worlds seems to be inappropriate.

　　There is an alternative view one might take of this semantic paradigm. As a purely formal notion it offers a rich model-theoretic tool for articulating and investigating the various systems of modal logic. This is the primary role it will play in this chapter and the next. In fact, the semantics for modal logic involves two notions: possible worlds and relations of accessibility between them. The relations of accessibility are meant to reflect the differences between the various modals. Formal constraints on this relation correspond to different modal axioms. In the present setting we have also to account for the truth predicate, which is now relativized to possible worlds. To provide the semantics of $L3$ our notion of model must contain all these components. The only difference between our notion and the standard one is the presence of the truth predicate. For each world in the model the truth (and falsity) predicate yields the extension of truth (falsity) at that world. Our models thus contain, in addition to all the standard apparatus of possible worlds, two functions from the domain of the model and the set of worlds to $\{0, 1\}$, which furnish the extensions of truth and modality. All this is packed into the following formal notion of model.

DEFINITION 6.0
A *model* for $L3$ is a structure $\mathcal{M} = \langle \mathcal{D}, W, R, T, F \rangle$ where

   (i) $\mathcal{D}$ is a model of the Lambda Calculus
   (ii) W is a set of *possible worlds*
   (iii) $R \subseteq W \times W$ is a relation of *accessibility*
   (iv) $T : D \times W \to \{0, 1\}$ and $F : D \times W \to \{0, 1\}$ are the extensions of truth and falsity.

The important difference between this and the original notion in chapter 2 should be clear: the truth and falsity predicates are now relativized to possible worlds. Observe that the domain of individuals is fixed by the Lambda Calculus model and we shall not consider models where the domain changes from world to world.

   The wff of the language $L3$ can now be given truth conditions in the standard way. Generally, we adopt pretty much the same notation as before but extended to allow for relativization to worlds. The intuitive picture which guides the semantics of the modal operator of necessity is given by the phrase: *A wff is necessarily true at some world w just in case it is true in all worlds accessible from w.* Consequently all the wff are evaluated relative to a world. More explicitly, we define $\mathcal{M} \vDash_{g,w} A$ by recursion on A as follows.

TRUTH CONDITIONS OF $L3$

$\mathcal{M} \vDash_{g,w} s = t$ iff $\mathcal{I}[t]_g = \mathcal{I}[s]_g$

$\mathcal{M} \vDash_{g,w} T(t)$ iff $T(\langle \mathcal{I}[t]_g, w \rangle) = 1$

$\mathcal{M} \vDash_{g,w} F(t)$ iff $F(\langle \mathcal{I}[t]_g, w \rangle) = 1$

$\mathcal{M} \vDash_{g,w} A \& B$ iff $\mathcal{M} \vDash_{g,w} A$ and $\mathcal{M} \vDash_{g,w} B$

$\mathcal{M} \vDash_{g,w} A \lor B$ iff $\mathcal{M} \vDash_{g,w} A$ or $\mathcal{M} \vDash_{g,w} B$

$\mathcal{M} \vDash_{g,w} A \to B$ iff $\mathcal{M} \vDash_{g,w} A$ implies $\mathcal{M} \vDash_{g,w} B$

$\mathcal{M} \vDash_{g,w} \sim A$ iff not $\mathcal{M} \vDash_{g,w} A$

$\mathcal{M} \vDash_{g,w} \forall x A$ iff for all d in D, $\mathcal{M} \vDash_{g(d/x),w} A$

$\mathcal{M} \vDash_{g,w} \exists x A$ iff for some d in D, $\mathcal{M} \vDash_{g(d/x),w} A$

$$\mathcal{M} \vDash_{g,w} \Box A \qquad \text{iff for every } w' \in W \ (wRw' \text{ implies } \mathcal{M} \vDash_{g,w'} A)$$
$$\mathcal{M} \vDash_{g,w} \Diamond A \qquad \text{iff for some } w' \in W \ (wRw' \text{ and } \mathcal{M} \vDash_{g,w'} A)$$

It is worth stressing again that the denotation of terms is fixed across possible worlds and we ignore modal logics where the domain of individuals varies. We shall often write $[A]._{//,g}$ for $\{w : \mathcal{M} \vDash_{g,w} A\}$, *the set of worlds at which A is true relative to g*, and say that a wff is *true at a world w, relative to g, in a model $\mathcal{M}$* iff $w \in [A]._{//,g}$. This is all quite standard, as indeed is the following definition of validity.

DEFINITION 6.1

A wff A of *L3* is *valid in a model* $\mathcal{M} = \langle \mathcal{D}, W, R, T, F \rangle$ iff $[A]._{//,g} = W$, for all assignment functions g.

The different systems of modal logic are all consequences of imposing different constraints upon the relation of accessibility. We summarize the basic terminology in the following definition.

DEFINITION 6.2

If the relation of accessibility is arbitrary we shall call the model a **K-*model***; if the relation is such that for every world there is at least one accessible world we shall call the model a **D-*model*** (or serial); if the relation of accessibility is reflexive we shall call the model a **T-*model***; if it is reflexive and transitive we shall refer to **S4-*models***; if it is an equivalence relation we shall refer to **S5-*models***.

Corresponding to these various notions of model we obtain different notions of validity.

DEFINITION 6.3

Let Γ be some class of models. We shall say that A is Γ-*valid* if it is valid in all models in Γ.

We shall be most concerned with the various classes of models given by the above restrictions on the relations of accessibility and employ **K**-validity to refer to those wff valid in all **K**-models, etc.

This ends the basic model theoretic background. To complete the picture we now briefly review the various modal logics and connect them with the properties of the accessibility relation.

6.3 Modal Logics

In all the models we shall consider, the Barcan formulae will be valid:

bar $\forall x \Box A \rightarrow \Box \forall x A$

No doubt there is another dimension we might investigate where the Barcan
formulae fails and indeed for certain philosophical purposes this might be
desirable. However, we simplify things in order to concentrate on other issues.
   We shall also assume the duality of $\Box$ and $\Diamond$, namely

dual $\Diamond A \leftrightarrow \sim \Box \sim A$

All the logics we shall consider have these axioms. We shall not attempt to
debate the merits of the Barcan formulae but refer the reader to the literature
cited.

(i) The Logic K

The weakest *normal* system of modal logic (K) is characterized by models in
which R is an arbitrary relation (K-models). It is defined by the following
axiom and rule:

ip $\Box (A \rightarrow B) \rightarrow (\Box A \rightarrow \Box B)$
nec   If $\vdash A$ then $\vdash \Box A$

From the perspective of metaphysical modal logic these are hardly
controversial and form part of every standard system. The first theorem
summarizes the soundness and completeness of the logic K with respect to the
class of K-models.

THEOREM 6.4
A wff is valid in all K-models (K-validity) iff it is a theorem of the logic K.

We shall not provide the proof of this since it will take us too far afield. We
refer the reader to Hughes and Cresswell [1968] for the details.
   This is the weakest logic of modality we shall consider. We have, it is true,
considered weaker logics of truth but the reasons for their study were different;
we were trying to sketch the space of consistent logics. This meant that the
progression to stronger logics had to be carried out with care.

(ii) The Logic D

The first extension we pursue is the modal logic we refer to as D. This has the
characteristic axiom:

dis $\Box A \rightarrow \Diamond A$

This is usually taken to be the basic deontic axiom and seems intuitively sound under the interpretation of $\Box$ as *obligation* and $\Diamond$ as *permission*: presumably, if one is obliged to perform some task, then permission to perform it ought to follow in its wake. It also seems sound under certain notions of (rational) belief where it is inconsistent to believe a proposition and its negation. Formally, the dis axiom just blocks the possibility that both $\Box(A)$ and $\Box(\sim A)$ can be simultaneously true.

THEOREM 6.5
A wff is valid in the class of D-models iff it is a theorem of the logic **D**.

This is still a very weak logic of modality and certainly does not capture the intuitions underlying metaphysical necessity. We now turn to stronger and more familiar systems. The literature of modal logic is actually dominated by three modal logics — **T**, **S4** and **S5**. We briefly study these in order of increasing strength.

(iii) **The Logic T**

The first of the famous trio of modal logics is **T**. Its characteristic axiom is

r $\Box A \rightarrow A$

Our terminology is somewhat non-standard. We have not employed **T** as the name of the axiom but reserved it for the logic itself. This is to avoid confusion later. This is the first logic in the sequence which has any real claim to be a logic of necessity: if A is necessarily true then surely A is true. This seems intuitively undeniable: it is difficult to imagine any notion of *necessary truth* from which *truth* fails to follow. However, if $\Box$ is interpreted as *believes that*, axiom r is intuitively unsound. This logic is characterized as follows.

THEOREM 6.6
A wff is valid in all **T**-models (**T**-validity) iff it is a theorem of the logic **T**.

Modal logicians do not usually stop here. They move on to even stronger systems. The fact that r corresponds to the relation of accessibility being reflexive suggests we move on to examine the implications of further constraints.

(iv) The Logic S4

The second member of the trio is the logic **S4**. This is the extension of **T** which is achieved by the addition of axiom:

tr $\Box A \rightarrow \Box\Box A$

The intuitive reasoning behind this axiom is more difficult to tease out. It is argued that what is necessary is necessarily so; perhaps so. **S4** is often taken to be the (epistemic) logic of knowledge. Under this interpretation $\Box$ is read as *knows that* and tr is justified by the observation that an agent who knows a proposition, knows that she knows it. It is also worth noting that tr is not implausible under the interpretation of $\Box$ as *believes that*. Rather than reflect upon this further, we move on to the soundness and completeness of **S4**.

THEOREM 6.7
A wff is valid in all **S4**-models (**S4**-validity) iff it is a theorem of the logic **S4**.

(v) The Logic S5

The final member of the trio is the logic **S5**. This is the extension of **S4** which is obtained by addition of the axiom:

e $\Diamond A \rightarrow \Box \Diamond A$

Under the interpretation of $\Box$ as *necessary that* and $\Diamond$ as *possible that* this might be justified but we shall not attempt to offer much intuitive justification for this axiom. Suffice it to say that under certain metaphysical notions of necessity what is possible is taken to be necessarily so. This logic also has its characterization theorem.

THEOREM 6.8
A wff is valid in all **S5**-models (**S5**-validity) iff it is a theorem of the logic **S5**.

The details of all the proofs for these theorems can be found in the literature cited. The axioms correspond exactly to each of the impositions on the relation of accessibility. We have not bothered to offer too much by way of intuitive justification for choosing between them but this is perhaps not the important point. The striking thing about the possible world semantics is that it provides a precise formal and metamathematical tool for articulating the differences between the possible logics. For the present this is all that concerns us.

    This completes our review of the standard account of modal logic. It should be sufficient background for what follows but the reader who requires more information should consult Hughes and Cresswell [1968]. We are now in a position to explore the modal extensions of each of the theories of truth elaborated in chapters 3, 4 and 5.

7 Truth in Modal Logic

So far we have only investigated theories of truth for the Classical Predicate Calculus. In this chapter we extend the development of the previous three chapters to Classical Modal Logic. This is not entirely for formal reasons. It is partly to pave the way to the second approach to modality which we take up in the next chapter. But this is not the whole story. The concept of necessary truth is the fundamental notion of modal logic and our main purpose in this chapter is to investigate the interaction between the operators of truth and modality.

The logics of the last chapter, although containing the truth predicate as part of their languages, contained no logic of truth since we imposed no axioms or rules for truth. Chapters 3, 4 and 5, however, supply us with an embarrassment of options. In this chapter we explore each of these three approaches within the setting of modal logic. As the reader will appreciate, the whole framework is much richer than before since we have to pay attention to the interaction of the operators of truth and modality. Despite this the results of the previous chapters all generalize quite smoothly.

7.1 Modal Fixpoints

We first consider the theory of truth studied in chapter 3 and develop its extension to the language of modal logic. The theory **KFG** is based upon the Kleene Truth conditions; our objective is to extend the theory to the language $L3$ of classical modal logic.

As with the theories of truth for the Predicate Calculus the axioms:

DIS $\quad \sim (T(t) \,\&\, F(t))$

FT $\quad F(A) \leftrightarrow T(\sim A)$

are taken to hold in all the theories we investigate. Syntactically things become a little messy and we shall employ parentheses freely to disambiguate wff.

(i) Modal Kleene Theories

The additional axioms are entirely predictable and simply extend the clauses of the Kleene truth conditions to the modal operators. More precisely, the clauses F1–F12 are extended by F13–F16, given below:

F13 $T(\Box A) \leftrightarrow \Box T(A)$
F14 $F(\Box A) \leftrightarrow \Diamond F(A)$
F15 $T(\Diamond A) \leftrightarrow \Diamond T(A)$
F16 $F(\Diamond A) \leftrightarrow \Box F(A)$

These new axioms reflect the interaction of truth and modality within the Kleene environment. For example, F13 can be read as "It is true that A is necessary iff it is necessary that A is true". Notice that they are independent of the choice of the underlying logic of modality. Indeed, there are many different theories buried here since the logic of modality is itself a possible parameter. Let Γ be any of the modal logics **K, D, T, S4** or **S5** and then let **KFGΓ** be the *Logic of Truth and Modality* which has as axioms those of **KFG** plus F13–F16, together with the axioms and rules for the modal logic Γ. So for example, **KFGK** is the logic which consists of F1–F16 plus:

bar $\forall x \Box A \rightarrow \Box (\forall x A)$
dual $\Diamond A \leftrightarrow \sim \Box (\sim A)$
ip $\Box (A \rightarrow B) \rightarrow (\Box A \rightarrow \Box B)$
nec If **KFGK** $\vdash$ A then **KFGK** $\vdash \Box A$

The others are obtained by the addition of the appropriate axioms of modality. So, for example **KFGT** is obtained by the addition of the T-axiom $\Box A \rightarrow A$; and **KFGS4** by the addition of the S4-axiom, etc. These are *modal theories of truth* and they reflect the formulation of Kleene semantics within the modal setting.

(ii) Models

Models of the theory are built in a similar fashion to those for **KFG** itself. The construction parallels that of the Predicate Calculus but everything is generalized to the modal case. There are several stages to the construction and we follow the pattern established in chapter 3.

We first provide the *Kleene semantics* for the language of modal logic. As with the non-modal language, we have to specify both truth and falsity conditions. The clauses for the logical connectives and quantifiers are simple generalizations of those for L but in addition, we must add the clauses for the modal connectives.

Let $\mathcal{M} = \langle \mathcal{D}, W, R, T, F \rangle$ be a model for $L3$. Define $\mathcal{M} \vdash_{w,g} A$ (A *is true at* w) and $\mathcal{M} \dashv_{w,g} A$ (A *is false at* w) by simultaneous recursion on A:

THE STRONG KLEENE TRUTH CONDITIONS FOR $L3$

$\mathcal{M} \vdash_{g,w} s = t$ iff $\mathcal{I}[t]_g = \mathcal{I}[s]_g$

$\mathcal{M} \vdash_{g,w} T(t)$ iff $T(\mathcal{I}[t]_g, w) = 1$

$\mathcal{M} \vdash_{g,w} F(t)$ iff $F(\mathcal{I}[t]_g, w) = 1$

$\mathcal{M} \vdash_{g,w} A \,\&\, B$ iff $\mathcal{M} \vdash_{g,w} A$ and $\mathcal{M} \vdash_{g,w} B$

$\mathcal{M} \vdash_{g,w} A \vee B$ iff $\mathcal{M} \vdash_{g,w} A$ or $\mathcal{M} \vdash_{g,w} B$

$\mathcal{M} \vdash_{g,w} A \rightarrow B$ iff $\mathcal{M} \dashv_{g,w} A$ or $\mathcal{M} \vdash_{g,w} B$

$\mathcal{M} \vdash_{g,w} \sim A$ iff $\mathcal{M} \dashv_{g,w} A$

$\mathcal{M} \vdash_{g,w} \forall x A$ iff for all d in D, $\mathcal{M} \vdash_{g(d/x),w} A$

$\mathcal{M} \vdash_{g,w} \exists x A$ iff for some d in D, $\mathcal{M} \vdash_{g(d/x),w} A$

$\mathcal{M} \vdash_{g,w} \Box A$ iff for all $w' \in W$ (wRw' implies $\mathcal{M} \vdash_{g,w'} A$)

$\mathcal{M} \vdash_{g,w} \Diamond A$ iff for some $w' \in W$ (wRw' and $\mathcal{M} \vdash_{g,w'} A$)

$\mathcal{M} \dashv_{g,w} s = t$ iff $\mathcal{I}[t]_g \neq \mathcal{I}[s]_g$

$\mathcal{M} \dashv_{g,w} T(t)$ iff $F(\mathcal{I}[t]_g, w) = 1$

$\mathcal{M} \dashv_{g,w} F(t)$ iff $T(\mathcal{I}[t]_g, w) = 1$

$\mathcal{M} \dashv_{g,w} A \,\&\, B$ iff $\mathcal{M} \dashv_{g,w} A$ or $\mathcal{M} \dashv_{g,w} B$

$\mathcal{M} \dashv_{g,w} A \vee B$ iff $\mathcal{M} \dashv_{g,w} A$ and $\mathcal{M} \dashv_{g,w} B$

$\mathcal{M} \dashv_{g,w} A \rightarrow B$ iff $\mathcal{M} \vdash_{g,w} A$ and $\mathcal{M} \dashv_{g,w} B$

$\mathcal{M} \dashv_{g,w} \sim A$ iff $\mathcal{M} \vdash_{g,w} A$

$\mathcal{M} \dashv_{g,w} \forall x A$ iff for some d in D, $\mathcal{M} \dashv_{g(d/x),w} A$

$\mathcal{M} \dashv_{g,w} \exists x A$ iff for all d in D, $\mathcal{M} \dashv_{g(d/x),w} A$

$\mathcal{M} \dashv_{g,w} \Box A$ iff for some $w' \in W$ (wRw' and $\mathcal{M} \dashv_{g,w'} A$)

$\mathcal{M} \dashv_{g,w} \Diamond A$ iff for all $w' \in W$ (wRw' implies $\mathcal{M} \dashv_{g,w'} A$)

This definition simply shadows the axioms of the theory given by F1–F16. We leave the reader to ponder the definition; we comment only on the new clauses. The clauses for necessity bifurcate into the conditions under which $\Box A$ is true and those under which it is false and so these truth conditions exactly reflect axioms F13–F16.

To construct a model for any of the modal theories we need to repeat the fixpoint construction of chapter 3. In order to set things up we require the definitions of *model extension* and of *Kleene revision*.

DEFINITION 7.0

Let $\mathcal{M} = \langle \mathcal{D}, W, R, T, F \rangle$ and $\mathcal{N} = \langle \mathcal{D}, W, R, T^*, F^* \rangle$ be two models for $L3$. Define

$\mathscr{M} \subseteq \mathscr{N}$ iff $\forall d \in D \forall w \in W(T(d, w) = 1$ implies $T^*(d, w) = 1)$
and $\forall d \in D \forall w \in W(F(d, w) = 1$ implies $F^*(d, w) = 1)$.

This is similar to the corresponding definition for L. The only difference concerns the possible worlds. In order to arrive at a model for F1–F16 we proceed as before and Kleene revise the extensions of truth and falsity. Let $\mathscr{M} = \langle \mathscr{D}, W, R, T, F \rangle$ be any model of $L3$.

We then define $\mathscr{M}' = \langle \mathscr{D}, W, R, T', F' \rangle$, the *Kleene revision of $\mathscr{M}$*, by

$$T'(\mathscr{I}[A]_g, w) = 1 \quad \text{iff} \quad \mathscr{M} \vdash_{g, w} A$$
$$F'(\mathscr{I}[A]_g, w) = 1 \quad \text{iff} \quad \mathscr{M} \dashv_{g, w} A$$

We first establish that the monotonicity property of the Kleene truth-conditions extends to the modal case. This is the content of the following.

THEOREM 7.1
Let $\mathscr{M}$ and $\mathscr{N}$ be models of $L3$. Then $\mathscr{M} \subseteq \mathscr{N}$ implies $\mathscr{M}' \subseteq \mathscr{N}'$.

PROOF By induction of the wff of $L3$. The two new cases involve $\square$ and $\diamond$. We illustrate with $\square$. Suppose that $\mathscr{M} \vdash_{g, w} \square A$. Then $\forall w'(wRw'$ implies $\mathscr{M} \vdash_{g, w'} A$). By induction $\forall w'(wRw'$ implies $\mathscr{N} \vdash_{g, w'} A$), as required. ●

Using this basic step of revision we proceed to construct a sequence of models. Apart from the fact that the extensions of truth and falsity are relativized to worlds, the definition is identical to that of chapter 3.

$$T(0) = \mathbb{0} \quad \text{where } (\forall d \in D \forall w \in W)(\mathbb{0}(d, w) = 0)$$
$$F(0) = \mathbb{0}$$
$$T(\alpha + 1) = T(\alpha)'$$
$$F(\alpha + 1) = F(\alpha)'$$
$$T(\delta)(d, w) = 1 \quad \text{iff } (\exists \alpha < \delta)(\forall \beta)(\alpha \leqslant \beta < \delta)(T(\beta)(d, w) = 1) \quad \text{(for } \delta, \text{ a limit}$$
$$F(\delta)(d, w) = 1 \quad \text{iff } (\exists \alpha < \delta)(\forall \beta)(\alpha \leqslant \beta < \delta)(F(\beta)(d, w) = 1) \quad \text{ordinal}$$

The mathematical tractability of this process of revision is again demonstrated by the fixpoint property, which is established exactly as in the non-modal case.

THEOREM 7.2
There is a model $\mathscr{M}^* = \langle \mathscr{D}, W, R, T^*, F^* \rangle$ such that $\mathscr{M}^* = (\mathscr{M}^*)'$.

As a corollary to this we have the existence of models for each of the modal theories under scrutiny.

THEOREM 7.3

Let $\mathcal{M} = \langle \mathcal{D}, W, R, T, F \rangle$ be a model for $L3$. Then if R is arbitrary, $\mathcal{M}^*$ is a model of **KFGK**; if every world has at least one accessible world then $\mathcal{M}^*$ is a model for **KFGD**; if R is reflexive then $\mathcal{M}^*$ is a model of **KFGT**; if R is reflexive and transitive then $\mathcal{M}^*$ is a model of **KFGS4**; and if R is an equivalence relation then $\mathcal{M}^*$ is a model of **KFGS5**.

PROOF It is clear from the discussion in chapter 3 that $\mathcal{M}^*$ is a model of **KFG**. The axioms F13–F16 follow directly from the extended Kleene truth conditions and the fact that the extensions of truth and falsity (internally) correspond to Kleene truth and falsity in the model. That the various modal logics correspond to the impositions placed upon the accessibility relation is clear from the classical correspondence theory of modal logic. ●

Theorem 7.3 reflects the fact that the theory **KFG** remains intact under the modal transformation. Moreover, the different modal extensions correspond exactly to the classical constraints on the relation of accessibility.

(iii) **A Reformulation**

We now examine the reformulation which arises from the Feferman–Gilmore presentation. Once again, the theories can be expressed in terms of the transformations $(+)$ and $(-)$ where these are extended to $L3$ as follows:

$$(\Box A)^+ = \Box(A^+)$$
$$(\Box A)^- = \Diamond(A^-)$$
$$(\Diamond A)^+ = \Diamond(A^+)$$
$$(\Diamond A)^- = \Box(A^-)$$

The theory based on C1 and C2 is defined exactly as before by the following axiom schemata:

C1 $T(A) \leftrightarrow A^+$
C2 $F(A) \leftrightarrow A^-$

We leave the reader to prove the following standard lemma.

LEMMA 7.4

(i) $A^+ \to A \;\&\; A^- \to \sim A$
(ii) $(A \leftrightarrow A^+) \to (T(A) \leftrightarrow A)$
(iii) $(\sim A \leftrightarrow A^-) \to (F(A) \leftrightarrow \sim A)$
(iv) $((A^+ \leftrightarrow A) \;\&\; (A^- \leftrightarrow \sim A)) \leftrightarrow P(A)$

The next theorem is again established as in the non-modal case. The additional clauses for the modal connectives induce no further complications.

THEOREM 7.5
The theory (A1−A6) + DIS + (F1−F16) is equivalent to DIS + (C1−C2), in the context of any of the modal logics, **K, D, T, S4** and **S5**.

In conclusion, the theory **KFG** extends uniformly to the modal context. Moreover, all the standard systems of modal logic are available to us and constitute a family of logics of truth and modality delineated by the properties imposed upon the relation of accessibility. This is important since the classical possible world semantics furnishes conceptual and mathematical insights which are not to be put aside lightly. We shall return to this point in chapter 8 where we investigate the second approach to modality.

These theories are formally very attractive. Like their non-modal parent they are easy to state and simple to apply. However, they inherit the curious features of **KFG** itself. If anything, their three-valued foundation seems even more curious in the context of modal logic — especially under the metaphysical interpretations of modality.

7.2   Modality and Stability

We next examine the approach to truth explored in chapter 4. What happens to the theory of stable truth in the context of modal logic? Here matters are much more involved since we have two systems of modal logic to contend with: the modal logic of truth and the actual logic of modality. We initiate the investigation by formulating the Gupta–Herzberger account of revision within the context of modal logic.

In the setting of $L3$, the definition of revision is a little more complex. Let $\mathcal{M} = \langle \mathcal{D}, W, R, T, F \rangle$ be a model for $L3$. Define the *Tarski revision* of $\mathcal{M}$, as $\mathcal{M}' = \langle \mathcal{D}, W, R, T', F' \rangle$, where

$$T'(\mathcal{I}[A]_g, w) = 1 \quad \text{iff} \quad \mathcal{M} \vDash_{g,w} A$$
$$F'(\mathcal{I}[A]_g, w) = 1 \quad \text{iff} \quad \mathcal{M} \vDash_{g,w} \sim A$$

The extension of truth is relativized to possible worlds and subsequently a pair $\langle \mathcal{I}[A]_g, w \rangle$ is in the new extension of truth just in case w is in the extension of A at the old model; otherwise it will be in the extension of falsity.

The concepts and results of the Gupta–Herzberger theory of revision apply in this context just as in the non-modal case. The definitions of *safeness*, *stability* and *stabilization model* are similar to the non-modal case. There are, however, some additional complications which stem from the class of models under consideration: depending on the class of models we get different notions of safeness and stability.

DEFINITION 7.6

Let Γ be any class of models for $L3$. Then we shall say that a wff A of $L3$ is Γ-*safe* iff A is valid at all stabilization models, for all initial models $\mathcal{M}$ in the class Γ.

It is Γ-*stably true* if T(A) is Γ-safe; it is Γ-*stably false* if F(A) is Γ-safe. By Γ-stability we mean Γ-stably true or Γ-stably false.

We now investigate the logics of safeness and stability and indicate how these logics are affected by the logics of modality. The whole theory generalizes in a predictable way. We shall not elaborate the theory in the same degree of detail as the development in chapter 4 since everything extends in a fairly uniform way. However, there will be points where some care needs to be taken in generalizing matters and here we shall explore things in more detail.

There are two degrees of freedom in the present setting corresponding to the actual modal logic and the logic of truth. There are so many options that it is quite difficult to know where to begin. Certain combinations, however, have a more natural structure. In particular, those in which the two logics are indentical seem formally more natural. Unfortunately, it will prove too restrictive just to study such combinations given the impositions that need to be placed upon the logics of truth. Nevertheless, for pedagogical reasons we begin with a logic in which the two logics coincide.

(i) **The Logic** D

We begin with a logic in which the logics of truth and modality are identical: in the logic D both the logic of truth and the logic of modality are **D**.

The logic consists of two groups of axioms and rules: one set for truth and one for modality. In order to avoid confusion we shall use lower case names for the axioms of modality and upper case for truth. It should also be stressed that there is a difference between the two operators. The modal operator is sentential whereas the truth operator is a one-place predicate. Although this difference does not affect the form of the axioms there is a conceptual difference of some importance which is more easily forgotten when the logics of both operators have a modal form.

        THE LOGIC D
DIS $T(A) \rightarrow C(A)$
IP $T(A \rightarrow B) \rightarrow (T(A) \rightarrow T(B))$

BAR $\forall x T(A) \rightarrow T(\forall x A)$

NEC If $D \vdash A$ then $D \vdash T(A)$

dis $\Box(A) \rightarrow \Diamond(A)$

ip $\Box(A \rightarrow B) \rightarrow (\Box(A) \rightarrow \Box(B))$

bar $\forall x \Box(A) \rightarrow \Box(\forall x A)$

nec   If $D \vdash A$ then $D \vdash \Box(A)$

We first establish the stability of this logic.

THEOREM 7.7

If $D \vdash A$ then A is **D**-stably true.

PROOF   We employ induction on the proofs in D. The proof for the axioms of truth follow the same pattern as the non-modal case. The axioms for $\Box$ are D-stably true since they are true at every world for every D-model. For NEC we have only to observe that if A is true at every world in every model from some ordinal onwards, then T(A) will be. For nec we observe that if A is true at every world at every model from some ordinal onwards, then $\Box(A)$ will be. ●

From our results about the concept of truth we know that the task of strengthening this logic is going to be a subtle and delicate one. Indeed, it is here where the logics of truth and modality part company. To point our intuitions in the right direction observe that the following are consequences of D.

AXIOMS OF ITERATED TRUTH AND MODALITY

$T(T(A) \rightarrow C(A))$

$T(T(A \rightarrow B) \rightarrow (T(A) \rightarrow T(B)))$

$T(\forall x T(A) \rightarrow T(\forall x A))$

$T(\Box(A) \rightarrow \Diamond(A))$

$T(\Box(A \rightarrow B) \rightarrow (\Box(A) \rightarrow \Box(B)))$

$T(\forall x \Box(A) \rightarrow \Box(\forall x A))$

$\Box(T(A) \rightarrow C(A))$

$\Box(T(A \rightarrow B) \rightarrow (T(A) \rightarrow T(B)))$

$\Box(\forall x T(A) \rightarrow T(\forall x A))$

$\Box(\Box(A) \rightarrow \Diamond(A))$

$\Box(\Box(A \rightarrow B) \rightarrow (\Box(A) \rightarrow \Box(B)))$

$\Box(\forall x \Box(A) \rightarrow \Box(\forall x A))$

These state that the modal and truth-theoretic analogues, of all the axioms are derivable. As might be predicted from the results of chapter 4, this property will fail for certain axioms of truth which extend **D**. Indeed, the two degrees of freedom alluded to earlier correspond to further axioms of truth and modality. We first consider some additional modal axioms.

(ii)   **Additional Modal Axioms**

We shall only be concerned with the standard modal axioms for the logics **T**, **S4** and **S5**. Recall that these take the form:

$$\text{MODALITY } (\Psi)$$

r	$\Box(A) \to A$
tr	$\Box(A) \to \Box(\Box(A))$
e	$\Diamond(A) \to \Box(\Diamond(A))$

Naturally, the safeness and stability of these modal axioms depends on the class of models involved (i.e. the accessibility relation).

THEOREM 7.8
r, tr, e are Γ-safe where Γ is respectively **T**, **S4** and **S5**.

  PROOF   This is clear since the axioms will be true at all worlds in all models of the appropriate kind.   ●

It is also obvious that the modal and truth-theoretic analogues of the axioms of modality (Ψ) are safe in the appropriate models. This is the content of 7.9 below. Consider the following axioms:

$\Box$**r**	$\Box(\Box(A) \to A)$
Tr	$T(\Box(A) \to A)$
$\Box$**tr**	$\Box(\Box(A) \to \Box(\Box(A)))$
Ttr	$T(\Box(A) \to \Box(\Box(A)))$
$\Box$**e**	$\Box(\Diamond(A) \to \Box(\Diamond(A)))$
Te	$T(\Diamond(A) \to \Box(\Diamond(A)))$

We leave the reader to convince themselves of the following.

THEOREM 7.9
The axiom pairs $\Box$r and Tr, $\Box$tr and Ttr, $\Box$e and Te, are Γ-safe where Γ is respectively **T**, **S4** and **S5**. In particular, r, tr and e are Γ-stably true where Γ is respectively **T**, **S4** and **S5**.

All this is obvious and perhaps not very exciting but we are trying to locate the exact point where the two modal systems part company and to gain some insight into the new modal logics which arise. Let Y be any subset of the modal axioms Ψ. The above discussion can be summarized in terms of the following logics of truth and modality. Let $D[Y]$ be the logic whose axioms are those of $D + Y$ together with the rules:

N1   If $D[Y] \vdash A$ then $D[Y] \vdash \square(A)$

N2   If $D[Y] \vdash A$ then $D[Y] \vdash T(A)$

N1 is the rule of necessitation for modality whereas N2 is the rule for truth. Notice the only axioms for truth are those given by D itself; we have only strengthened the modal part of the theory. The stability of this logic follows from the above discussion.

THEOREM 7.10

If $D[Y] \vdash A$ then A is Γ-stably true where Y is $\{r\}$, $\{r, tr\}$ or $\{r, tr, e\}$ and Γ is respectively **T**, **S4** and **S5**.

PROOF   That the axioms are stably true follows from the stable truth of the D-axioms (7.7) and 7.9. For the rules we have only to observe that if A is true in all worlds from some ordinal onwards then both $\square(A)$ and $T(A)$ will be.   ●

This extension to D only involves the additional axioms of modality r, tr and e. As a consequence all the theorems are stably true. We now turn to the addition of further axioms of truth.

(iii)   **Stable Axioms of Truth: Logics with Full Necessitation**

Which additional axioms of truth should we consider? We take our lead from the discussion of chapter 4. To begin with we consider those axioms which we know to be stably true:

S $T[T(A) \rightarrow A] \rightarrow P(A)$

Q $T[T(A) \rightarrow T(T(A))] \rightarrow P(A)$

W $[C(A) \rightarrow T(C(A))] \rightarrow P(A)$

There are three different (families of) logics of truth and modality, of increasing strength, which can be obtained by adding these axioms to those of $D[Y]$. More precisely, let Y be some subset of the axioms of modality Ψ and then let $ST[Y]$ be the following logic.

THE LOGIC ST[Y]

DIS $T(A) \to C(A)$
IP $T(A \to B) \to (T(A) \to T(B))$
S $T(T(A) \to A) \to P(A)$
BAR $\forall x T(A) \to T(\forall x A)$
NEC If $ST[Y] \vdash A$ then $ST[Y] \vdash T(A)$

AXIOMS AND RULES OF MODALITY

The axioms of modality Y plus the following:
dis $\Box(A) \to \Diamond(A)$
ip $\Box(A \to B) \to (\Box(A) \to \Box(B))$
bar $\forall x \Box(A) \to \Box(\forall x A)$
nec If $ST[Y] \vdash A$ then $ST[Y] \vdash \Box(A)$

The logic SS4[Y] is obtained by the addition of Q, to the axioms of truth, and SS5[Y] by the further addition of W. The stability of these axioms of truth ensures that the whole logic is well-behaved. Notice that, once r is in place, dis is redundant but we shall not fuss over this.

THEOREM 7.11

(i) If $ST[Y] \vdash A$ then A is Γ-stably true where Y is $\{r\}$, $\{r, tr\}$ or $\{r, tr, e\}$ and Γ is respectively **T**, **S4** or **S5**.

(ii) If $SS4[Y] \vdash A$ then A is Γ-stably true where Y is $\{r\}$, $\{r, tr\}$ or $\{r, tr, e\}$ and Γ is respectively **T**, **S4** or **S5**.

(iii) If $SS5[Y] \vdash A$ then A is Γ-stably true where Y is $\{r\}$, $\{r, tr\}$ or $\{r, tr, e\}$ and Γ is respectively **T**, **S4** or **S5**.

PROOF   It is sufficient to establish (iii) since this is the strongest logic. The proof is identical to that of 7.10. Each of the axioms of truth is stably true with respect to all classes of models, and as before the stable truth of the new axioms of truth guarantees that the full rule of necessitation preserves stable truth. ●

These logics are the natural outcome of combining the standard modal logics with the stable logics of truth. They are all extensions of $D[Y]$ and arise by adding further stable axioms of truth to those of $D[Y]$. We now examine those axioms which are safe but unstable.

(iv) **Safe but Unstable Axioms of Truth**

We again turn to our previous discussion for inspiration. In chapter 4 we uncovered the following safe but unstable axioms:

$$\text{TRUTH } (\Upsilon)$$
R $T(A) \to A$
Tr $T(A) \to T(T(A))$
IPT $T(T(A) \to T(B)) \to T(A \to B)$

While we are reflecting upon these we might as well consider their modal analogues:

$\square\,[T(A) \to A]$
$\square\,[T(A) \to T(T(A))]$
$\square\,[T(T(A) \to T(B)) \to T(A \to B)]$

THEOREM 7.12
R, Tr, IPT and their modal analogues are Γ-safe where Γ is **T**, **S4** or **S5**.

    PROOF   Consider the first by way of an example. For this and its modal analogue to be Γ-safe it has to be true at every world for every stabilization model, and the argument for this is identical to the non-modal case.   ●

We can go further and formulate a logic which is fully cognizant of all of these facts. To facilitate the discussion we again require a little notation to distinguish the various logics. Let $ST[X; Y]$ be the logic whose axioms are those of $ST[Y]$ plus the safe axioms of truth X, where X is some subset of Υ ($= \{R, Tr, IPT\}$) together with the rules:

if $ST[X; Y] \vdash A$ then $ST[X; Y] \vdash \square\,(A)$
if $ST[Y] \vdash A$ then $ST[X; Y] \vdash T(A)$

where $ST[Y]$ is used in the sense of 7.11. Indeed, $ST[Y]$ is a special case of the above where the set X is empty. $SS4[X; Y]$ and $SS5[X; Y]$ are obtained in a parallel fashion from $SS4[Y]$ and $SS5[Y]$, respectively. With these logics we only obtain their safeness.

THEOREM 7.13

(i)   If $ST[X; Y] \vdash A$ then A is Γ-safe where Y is $\{r\}$, $\{r, tr\}$ or $\{r, tr, e\}$ and Γ is respectively **T**, **S4** or **S5**, and X is any combination of the safe axioms of truth.

(ii)   If $SS4[X; Y] \vdash A$ then A is Γ-safe where Y is $\{r\}$, $\{r, tr\}$ or $\{r, tr, e\}$ and Γ is respectively **T**, **S4** or **S5**, and X is any combination of the safe axioms of truth.

(iii)   If SS5 [X; Y] ⊢ A then A is Γ-safe where Y is {r}, {r, tr} or {r, tr, e} and Γ
is respectively T, S4 or S5, and X is any combination of the safe axioms
of truth.

PROOF   (i) and (ii) are special cases of (iii), so we concentrate on it. If A is
an instance of any of the axioms we employ 7.11 and 7.12. Consider the first
rule. If A is true at every world for every stabilization model then □ (A) will be.
The rule for truth follows from the fact that all the theorems of SS5 [Y] are
Γ-stable where Y is any combination of modal axioms.   ●

Observe that this result cannot be strengthened to include the full rule of
necessitation for the truth predicate since the axioms of truth, Υ, are safe but
unstable. Thus we seem to have pushed things as far as possible. However,
there is one further point which deserves attention. The next result links
modality and truth.

THEOREM 7.14
For all wff A, □ (T(A)) ↔ T(□ (A)) and ◇ (F(A)) → F(□ (A)) are Γ-safe where
Γ is K, D, T, S4 or S5.

PROOF   For the first let σ be any stabilization ordinal. The left-hand side
amounts to

$$\forall w \forall w' [R(w, w') \to (\forall \beta \geqslant \sigma) \mathcal{M}(\beta) \vDash_{g, w'} A]$$

and the right-hand side to

$$\forall w \forall \beta \geqslant \sigma \forall w' [R(w, w') \to \mathcal{M}(\beta) \vDash_{g, w'} A]$$

which are clearly equivalent. The second is left as an exercise.   ●

The reader might consider which other parts of F13–F16 are safe.
It is not difficult to see that the modal analogues of the above are also safe.

THEOREM 7.15
For all wff A, □ [T(□ (A)) ↔ □ (T(A))] and □ [◇ (F(A)) → F(□ (A))] are
Γ-safe where Γ is K, D, T, S4 or S5.

It appears that the truth-theoretic analogue of this axiom is not stably true;
actually I have been unable to establish the matter one way or the other.
However, these last two results inform us that we can include the axioms

T□   T(□ (A)) ↔ □ (T(A))
       ◇ (F(A)) → F(□ (A))

in our modal logics as new safe axioms of truth.

Let $ST'[X; Y]$ be $ST[X + \{T\square\}; Y]$. $SS4'[X; Y]$ and $SS5'[X; Y]$ are obtained from $SS4[X; Y]$ and $SS5[X; Y]$ in a parallel fashion. In what follows we shall abbreviate $ST'[X; Y]$ as $\Gamma T[X]$, where it is understood that Γ is the modal logic whose axioms correspond to those of T, $S4$ or $S5$. Similarly, we write $\Gamma S4[X]$ for $SS4'[X; Y]$ and $\Gamma S5[X]$ for $SS5'[X; Y]$, respectively. This notation enables us to bury the axioms of modality Y in Γ. Where X is empty we shall just write $\Gamma[T]$, $\Gamma[S4]$ or $\Gamma[S5]$.

This completes our discussion of the logics of stable truth and modality. Again, it is by no means a complete treatment but we trust that we have provided sufficient background to enable the reader to pursue the subject further. Perhaps the most natural systems are those $ST[Y]$, $SS4[Y]$ and $SS5[Y]$ which admit full principles of necessitation for both truth and modality. We recommend these for further study.

7.3   A Theory of Propositions, Truth and Modality

Finally, we consider the modal extensions of the theory of Frege Structures. Once more the theory extends quite elegantly to the modal setting. We are required to extend the axioms of a Frege Structure to include axioms of propositions and truth for the modal operators. We first enrich the language of Frege Structures.

(i)   The Language

The language is different from $L3$ since it has the language $L2$ as its non-modal part. Call this language $L4$. Because it is non-standard we provide the language in full. It has the following content.

BASIC VOCABULARY OF $L4$
Individual variables $x, y, z, \ldots$
Individual constants $c, d, e, \ldots$
Logical combinators $\vee, \wedge, \neg, \Rightarrow, \Leftrightarrow, \Xi, \Theta, \approx, \blacksquare$

The new logical combinator $\blacksquare$ is the internal or intensional modal. It is on a par with all the other intensional connectives and quantifiers.

INDUCTIVE DEFINITION OF TERMS
(i)   Every variable, combinator or constant is a term.
(ii)   If t is a term and x is a variable then $(\lambda x.t)$ is a term.
(iii)   If t and t' are terms then (tt') is a term.

INDUCTIVE DEFINITION OF WELL-FORMED FORMULAE

(i)   If t and s are terms then $s = t$, Prop(t) and True(t) are (atomic) wff.

(ii)   If Φ and Φ' are wff then $\Phi \& \Phi'$, $\Phi \vee \Phi'$, $\Phi \rightarrow \Phi'$, $\sim\Phi$ are wff.

(iii)   If Φ is a wff and x a variable then $\exists x\Phi$ and $\forall x\Phi$ are wff.

(iv)   If Φ is a wff then $\square\Phi$ is a wff.

All the rest of the syntax is identical to that of chapter 5. We have not included the dual modal operator but we can safely leave such an addition to the interested reader.

(ii)   The Theories

The axioms of the theory are those of a Frege Structure (see chapter 5) plus some new axioms for the modal connectives. The additional axioms govern the interaction of truth and modality:

(viii) $\square(\text{Prop}(t)) \rightarrow \text{Prop}(\blacksquare\, t)$

(ix) $\square(\text{Prop}(t)) \rightarrow (\text{True}(\blacksquare\, t) \leftrightarrow \square(\text{True}(t)))$

The axiom (ix) for truth is intuitively clear. The axiom for propositions parallels those for universal and existential quantification: for $\blacksquare$ A to be a proposition, A has to be a proposition in all accessible worlds, i.e. A is necessarily a proposition.

Let **FSK** be the theory which consists of the axioms (i)–(viii) for propositions and (i)–(ix) for truth, plus the following axioms and rule:

ip $\square(A \rightarrow B) \rightarrow (\square A \rightarrow \square B)$

bar $\forall x(\square A) \rightarrow \square(\forall xA)$

nec **FSK** $\vdash A$ then **FSK** $\vdash \square(A)$

This is the weakest normal system since we have only included the modal logic **K** as part of the theory. We might wish to consider the extensions generated by the other modal logics. Let **FSD** be the extension obtained by adding the axiom dis of modal logic; **FST** is obtained by addition of the r-axiom, **FSS4** by the addition of the tr-axiom and **FSS5** by the further addition of the e-axiom. We shall not consider logics weaker than **FSD**.

(iii)   Models — Extensions of TP

To build models of these systems within the fixpoint and stable theories we first consider the modal extensions of the theory **TP**. Let **TP**Γ be the theory cast within the language $L3$ which consists of the axioms of **TP** plus the modal axioms:

(a) $\Box(P(A)) \to P(\Box A)$

(b) $\Box(P(A)) \to (T(\Box A) \leftrightarrow \Box(T(A)))$

together with the axioms and rules of the modal logic Γ (for $\Gamma = D, T, S4, S5$).

THEOREM 7.16

TPΓ is derivable in both KFGΓ and Γ[S5].

PROOF   We establish this by induction on the proofs in TPΓ. Once again the proof is a simple extension of the non-modal case. We have only to concern ourselves with the axioms (a) and (b) and these are a direct consequence of F13–F16 in the case of KFGΓ and virtually by stipulation (using T$\Box$) in the case of Γ[S5]. Finally, the rules are automatic since the rules of TPΓ are also rules of the other two systems.   ●

With this result in place, the modal Frege Structures axioms are easily derivable. There is one preliminary step we must take, namely the definition of the intensional modal. This is accomplished as follows.

DEFINITION 7.17

$$(\blacksquare t) = _{\mathrm{def}} \hat{\ }(\Box T(t))$$

This parallels the definition of the other connectives and quantifiers in definition 5.0.

THEOREM 7.18

With the above definition together with those of 5.0, in the theory TPΓ, the axioms of the modal Frege Structure FSΓ are derivable for $\Gamma = D, T, S4, S5$.

PROOF   Most of the details are similar to the non-modal case. We have to check in addition that the axioms (viii) of propositions and (ix) for truth are sound under this interpretation and these are direct consequences of (a) and (b).   ●

This completes our exposition of modal Frege structures. Again most of the work has been routine and there is little doubt that a more detailed exposition would uncover more of interest. We shall, however, not pursue matters further.

In this chapter we have extended the three theories of truth to the modal context. We leave the reader to carry out further excavations. There are many details which remain to be filled in but most are routine and can be safely left to the reader as expositional exercises.

8 Predicative Modality

> *"The advantage of such a treatment is obvious: if modal terms become predicates, they will no longer give rise to non-extensional contexts, and the customary laws of the predicate calculus with identity may be employed"*
> Montague[1963]

The traditional approach to modal logic treats the modal operators as new one-place connectives on a par with negation, conjunction and the other propositional connectives. In this chapter we investigate the alternative approach where the modal operators are treated as predicates. This parallels the move taken with regard to the truth predicate, and represents a more unified approach to truth and modality and one that reflects the uniformity of natural language in regard to its representation of these concepts.

The advantages of this approach have already been discussed in chapter 1 but perhaps they bear repetition. They are put quite well in the quote from Montague: if modal operators are predicates then we shall not be plagued with the problem of non-extensional contexts and consequently we shall be able to reason about their arguments by employing ordinary first-order logic with equality. Unfortunately, the negative results of Montague and Kaplan[1960], Montague[1963] and Thomason[1980] on the intensional paradoxes suggest that such an approach is not viable. However, the work of the previous chapters suggests that there is hope for the development of useful theories.

8.1 The Language

The "syntactical" treatment of modality requires a different syntactic setting. We extend L in a different way to that of the previous chapter. Let $L5$ be the language which is obtained from L by the addition of the following clause:

If t is a term then $\blacktriangle (t)$ and $\blacktriangledown (t)$ are atomic wff.

We thus employ $\blacktriangle$ and $\blacktriangledown$ as the modal predicates where we abbreviate $\blacktriangle (\hat{}A)$ as $\blacktriangle (A)$, etc. The new language therefore has five kinds of atomic wff:

the assertions of truth, falsity, equality and the modalities. ▲ is meant to be the stronger of the two modals. The language is equipped with its semantics in the same way as L but now the models are enriched with two new functions $J: D \rightarrow \{0, 1\}$ and $I: D \rightarrow \{0, 1\}$, which provide the extension of the modal predicates ▲ and ▼. The models of $L5$ thus have the following general form: $\mathcal{M} = \langle \mathcal{D}, T, F, J, I \rangle$. In addition, we add new clauses to the semantics:

$$\mathcal{M} \vDash_g \text{▲} (A) \quad \text{iff} \quad J(\mathscr{I}[A]_g = 1$$
$$\mathcal{M} \vDash_g \text{▼} (A) \quad \text{iff} \quad I(\mathscr{I}[A]_g) = 1$$

The abstract form of the semantics thus causes no problems. The central question concerns the logic of the modals: which logics of ▲ and ▼ are available to us? Our experience with the truth predicate should give us pause for thought: if these logics take on the form of the standard logics T, $S4$ or $S5$ then problems arise immediately. If anything our situation is now worse since we have predicates of both truth and modality and paradox might arise through either one or indeed through a combination of both. The central concern of the present chapter is to uncover, in a principled semantical way, the consistent logics of truth and predicative modality. First we explicitly demonstrate how paradox arises when modality is treated predicatively.

8.2 Intensional Paradox

The message of the intensional paradoxes such as the *Knower* (Montague [1963]) is that the logic of the modal predicates must not be too strong. The *Knower* involves a self-referential sentence which asserts that its negation is known:

$$A \leftrightarrow \text{▲} (\sim A)$$

where here ▲ is to be interpreted as *knows that*. The sentence emerges by reflection upon the so-called *unexpected examination* or *knower paradox*. We give a brief account of this.

The situation is as follows. A teacher tells his class that they will be given an examination sometime during the following week. The teacher does not say on which day the examination will be; he wants it to be a surprise. Can the teacher achieve his goal of giving the class a surprise examination? One of the students decides that he cannot. She reasons as follows. The teacher cannot give the examination on Friday (the last working day of the week) since, if no examination has materialized by the time Friday morning arrives (indeed, Thursday evening), it would then not be a surprise. For similar reasons, the student rules out Thursday and subsequently all the other days of the week.

The student concludes that the teacher cannot set an examination which will come as a surprise: the teacher cannot carry out the two halves of his threat. Unfortunately for the class the examination occurs on Wednesday and all are surprised including the precocious student. This definitely has the air of paradox about it. We shall not take time to formalize this argument and uncover its logical structure. However, after a certain amount of refinement and reflection it is possible to discern the existence of the self-referential sentence:

$$A \leftrightarrow \blacktriangle (\sim A)$$

which as we indicated asserts that its negation is known. A detailed analysis (and a certain amount of refinement) also yields the following principles implicit in one formal representation of the argument.

I

r	$\blacktriangle (A) \to A$
ip	$\blacktriangle (A \to B) \to (\blacktriangle (A) \to \blacktriangle (B))$
a	$\blacktriangle (\blacktriangle (A) \to A)$
nec	If $\mathsf{LC} \vdash A$ then $\blacktriangle (A)$

This is certainly weaker than the modal logic **T** (which we know to be inconsistent) since, in **I**, nec is a restricted rule. It is arguable whether all or any of these principles are plausible under the interpretation of $\blacktriangle$ as *knows that*. Sainsbury [1988] contains further discussion on this issue and offers an illuminating discussion of the paradox. Despite the weakness of this logic, we have the following result essentially due to Montague [1963].

THEOREM 8.0
The modal logic **I** is inconsistent.

  PROOF   From the diagonalization lemma and nec we know there is a wff A such that

(a) $\blacktriangle (A \leftrightarrow \blacktriangle (\sim A))$

From the tautology $(A \leftrightarrow \sim A) \to \sim A$ and nec we obtain

(b) $\blacktriangle ((A \to \sim A) \to \sim A)$

From the tautology

$$[(A \to \sim A) \to \sim A] \to [(A \leftrightarrow \blacktriangle (\sim A)) \to ((\blacktriangle (\sim A) \to \sim A) \to \sim A)],$$

nec yields

(c)   ▲ ([[(A → ~A) → ~A] →

           [(A ↔ ▲ (~A)) → ((▲ (~A) → ~A) → ~A)]])

(a), (b) and (c) yield, by ip,

(d)   ▲ ((▲ (~A) → ~A) → ~A)

Employing axioms a and ip, from (d) we obtain

(e)   ▲ (~A)

By (e) and r we obtain: (f)   ~A
   On the other hand, by (a) and r we obtain: (g) A ↔ ▲ (~A)
which by (e) yields: (h)   A
   (f) and (h) are contradictory.   ●

This destroys any hope of any strong logic of the modal operator as a predicate. Of course, the same argument would have destroyed a logic of truth of the above form but here the situation seems far worse for we know that such logics are available for the modal operator treated as a connective rather than a predicate. On the face of it we have good reason to reject this whole notion of syntactic modality. However, not all is lost. The lesson to be gleaned from the contradiction is that we must be more circumspect in the way the logics are postulated. In the next two sections we shall address this issue, deriving our inspiration from the results of the previous chapter. Before we do so, however, we consider a logic which involves belief rather than knowledge. To indicate the intended interpretation we write bel for ▲ in this version.

J

tr   bel(A) → bel(bel(A))
dis   bel(A) → ~bel(~A)
imp bel(A → B) → (bel(A) → bel(B))
nec If J⊢ A then J⊢ bel(A)

THEOREM 8.1
J is inconsistent.

   PROOF By diagonalization we can locate a sentence A such that A ↔ ~bel(A). By nec we obtain immediately (1): bel(A ↔ ~bel(A)). Assume (2), namely bel(A). By tr we obtain (3): bel(bel(A)). By dis we have (4): bel(bel(A)) → ~bel(~bel(A)). By (3) and (4) we can conclude (5):

~ bel(~ bel(A)). By (1) and imp we are entitled to conclude (6): bel(A) → bel(~ bel(A)). By (5) and (6), we obtain (7): ~ bel(A). By (2), discharging the assumption, we obtain (8): bel(A) → ~ bel(A). Hence, (9): ~ bel(A). Using nec we can conclude (10): bel(~ bel(A)). By (1) again we can conclude (11): bel(A). Finally, (11) and (9) are contradictory. ●

These paradoxes suggest that developing logics of predicative modality is going to be no easy task. If this were not bad enough we have the additional difficulties created by the truth predicate itself and the interaction of the two predicates.

8.3 A Translation between *L5* and *L3*

Once again we appear to have reached an impasse. Our only way out seems to be to deny one or more of the principles of these logics. It is not hard to find reasons for doubting some or indeed all of the above principles especially under the interpretation of ▲ as *believes that*. Imp itself looks rather dubious in full generality although even here one might well be prepared to admit that the particular use made of this principle in the proof is intuitively justified. We are not, however, primarily concerned with the philosophical considerations which impinge on the above principles. Our task is a more general one, namely to explore the logical space of possibilities: which logics are available to us? However, this is not just a question of consistency. We cannot circumvent this problem in any theoretically satisfactory manner by merely postulating weaker logics. It is easy enough to find weak and useless logics which are consistent. At the very least we must demand some semantic account of their origin. Fortunately, the theories of truth and modality expressed in *L3* can be marshalled to provide some insight. To facilitate this we proceed in a slightly different way to the previous chapters. Instead of treating ▲ and ▼ as new predicates we can employ the classical modal connectives and the truth predicate to define these modal predicates. Indeed, in the context of *L3*, the modal predicates admit of obvious definitions:

$$\blacktriangle (t) =_{\text{def}} \square (T(t))$$
$$\blacktriangledown (t) =_{\text{def}} \lozenge (T(t))$$

The logic of the modal predicate can then be unpacked in a semantic way using the logics of truth and the logics of classical modality. However, this is not quite the route we shall adopt. Rather we shall use the above as a basis for a translation between the language of syntactic modality *L5* and the language of classical modal logic *L3*. As with classical modal logic, the semantic theory of *L3* (together with the logics of truth) provides us with a semantic way of articulating the differences between the different modal logics. In this section we shall be guided by the account given in section 7.2. Kamp and Ascher [1988] have also pursued the stability approach to the semantics of predicative modality along similar lines. They interpret ▲ directly and provide a Gupta–Herzberger revision theory of modality. They have not, however, been too concerned with the explicit formulation of the underlying logics. Their paper is more concerned with a detailed analysis of various forms of self-reference. We recommend it to the interested reader.

This proposed translation facilitates the lifting of the semantic insights from $L3$ to $L5$. More explicitly, the translation from $L5$ to $L3$ is given by the following recursion:

(i) $\tau(x)$ $= x$
(ii) $\tau(\hat{}A)$ $= \hat{}\tau(A)$
(iii) $\tau(tt')$ $= \tau(t)\tau(t')$
(iv) $\tau(\lambda x.t)$ $= \lambda x.\tau(t)$
(v) $\tau(▲(t))$ $= \Box(T(\tau(t)))$
(vi) $\tau(▼(t))$ $= \Diamond(T(\tau(t)))$
(vii) $\tau(T(t))$ $= T(\tau(t))$
(viii) $\tau(A \& B) = \tau(A) \& \tau(B)$
(ix) $\tau(\sim A)$ $= \sim \tau(A)$
(x) $\tau(\forall xA)$ $= \forall x\tau(A)$

The clauses for the other connectives and quantifiers are obtained by the standard definitions. The important clauses are (v) and (vi); the others are routine. This translation provides the technical means of investigating our logics of predicative truth and modality by employing the semantic settings of possible worlds and stable truth. We are thus able to articulate the differences between the various modal logics of predicative modality in terms of the properties of the relation of accessibility and the concept of stability. Our notion of predicative modality is thus filtered through the truth predicate and sentential modality.

LOGIC OF		LOGIC OF		LOGIC	
PREDICATIVE	=	SENTENTIAL	⊕	OF	
MODALITY		MODALITY		TRUTH	

8.4   Logics of Truth and Modality with Full Necessitation

We are now in a position to explore the various logics of truth and modality which are available to us within the setting of $L5$. We initiate the discussion with the logic **D**, which parallels the logic D of chapter 7.

THE LOGIC OF TRUTH AND MODALITY D

AXIOMS AND RULES FOR TRUTH
DIS $T(A) \rightarrow \sim T(\sim A)$
IP $T(A \rightarrow B) \rightarrow T(A) \rightarrow T(B)$
BAR $\forall x T(A) \rightarrow T(\forall x A)$
NEC If $D \vdash A$ then $D \vdash T(A)$

AXIOMS AND RULES OF MODALITY
dis $\blacktriangle (A) \rightarrow \sim \blacktriangle (\sim A)$
ip $\blacktriangle (A \rightarrow B) \rightarrow (\blacktriangle (A) \rightarrow \blacktriangle (B))$
bar $\forall x \blacktriangle (A) \rightarrow \blacktriangle (\forall x A)$
nec If $D \vdash A$ then $D \vdash \blacktriangle (A)$

THEOREM 8.2
The logic **D** is a consistent logic of truth and predicative modality.

PROOF   To show consistency we employ the translation and the logic D of truth and modality of $L3$. First consider dis. Under the translation we have to show that

$$\sim (\blacktriangle (A) \& \blacktriangle (\sim A)) \quad \text{i.e.} \quad \sim (\square (T(A)) \& \square (T(\sim A)))$$

Suppose that $\square (T(A)) \& \square (T(\sim A))$. Using the DIS axiom for T we get $T(\sim A) \rightarrow \sim T(A)$; from the nec axiom for $\square$ we obtain $\square (T(\sim A) \rightarrow \sim T(A))$. From the assumptions and ip for $\square$ we can deduce $\square (\sim T(A))$. But now we have $\square (T(A)) \& \square (\sim T(A))$ which contradicts dis for $\square$.

The axiom for detachment $\blacktriangle (A \rightarrow B) \& \blacktriangle (A) \rightarrow \blacktriangle (B)$ translates to

$$\square (T(A \rightarrow B)) \& \square (T(A)) \rightarrow \square (T(B))$$

Here we employ ip, nec for $\square$ and ip for $\square$ twice. Barcan is also straightforward: we employ Barcan for T, nec for $\square$, Barcan for $\square$, and finally nec for $\square$. Finally consider the rule of necessitation itself. Here we employ NEC for T followed by nec for $\square$.   ●

Perhaps this is not too surprising given the results of the previous chapter. The modal part of the theory caused no problems; we can have pretty much what we want. The constraints all come from the logic of truth. Given that the predicative modal is defined essentially in terms of the sentential modal and the truth predicate, it is the logic of the latter which constrains the possibilities.

The obvious extensions to these logics involve the axioms of truth and modality from the previous chapter. Since here we are considering full necessitation we begin with the stably true axioms of truth.

AXIOMS OF STABLE TRUTH
S $T[T(A) \rightarrow A] \rightarrow P(A)$
Q $T[T(T(A)) \rightarrow T(A)] \rightarrow P(A)$
W $[C(A) \rightarrow T(C(A))] \rightarrow P(A)$

Using these additional stable axioms of truth we are at liberty to formulate a family of logics of truth and predicative modality. Let **ST** be the following logic:

ST

AXIOMS AND RULES FOR TRUTH
DIS $T(A) \rightarrow \sim T(\sim A)$
IP $T(A \rightarrow B) \rightarrow (T(A) \rightarrow T(B))$
BAR $\forall x T(A) \rightarrow T(\forall x A)$
S $T[T(A) \rightarrow A] \rightarrow P(A)$
NEC If **ST** $\vdash$ A then **ST** $\vdash$ T(A)

AXIOMS AND RULES FOR MODALITY
dis $\blacktriangle (A) \rightarrow \sim \blacktriangle (\sim A)$
ip $\blacktriangle (A \rightarrow B) \rightarrow (\blacktriangle (A) \rightarrow \blacktriangle (B))$
bar $\forall x \blacktriangle (A) \rightarrow \blacktriangle (\forall x A)$
nec If **ST** $\vdash$ A then **ST** $\vdash \blacktriangle (A)$

The logics **SS4** and **SS5** are obtained by the further additions of Q and W, respectively. The proof of the following result follows the same line of argument as the previous one.

THEOREM 8.3
The logics **ST**, **SS4** and **SS5** are consistent.

These logics are all well-behaved but the next extensions are not. Indeed, as will become clear, the extensions are much less well-behaved than the corresponding logics of the previous chapter, since here the problems which plague the truth predicate also infect the modal operators.

8.5 Logics of Truth and Modality without Full Necessitation

These extensions involve the axioms of modality for T and S4 modal logic and the safe but unstable axioms of truth. From here onwards we must proceed with great care since the modal axioms themselves are no longer immune from instability. We first reformulate the axioms in the context of *L*5.

MODALITY (Ψ)

r $\blacktriangle (A) \rightarrow A$

tr $\blacktriangle (A) \rightarrow \blacktriangle (\blacktriangle (A))$

Observe that the e-axiom is missing in the list of modal axioms. The omission results from the fact that e is not safe in the revision process. We again remind the reader of the safe but unstable axioms of truth.

SAFE AXIOMS OF TRUTH (Υ)

R $T(A) \rightarrow A$

Tr $T(A) \rightarrow T(T(A))$

IPT $T(T(A) \rightarrow T(B)) \rightarrow T(A \rightarrow B)$

In the context of *L*5, let **ST** [X; Y] be the logic whose axioms are those of **ST** plus the axioms of truth X and the axioms of modality Y (where X are the axioms of truth selected from Υ and Y are the axioms of modality selected from Ψ), together with the rules:

ST $\vdash$ A then **ST** [X; Y] $\vdash$ $\blacktriangle$ (A)

ST $\vdash$ A then **ST** [X; Y] $\vdash$ T(A)

Observe that the rules of necessitation are restricted to inferences involving the logic **ST**. This is a consequence of the fact that the modal predicate is defined through the truth predicate and so the restrictions for truth apply to modals as well. This is to be seen in contrast to the approach of the previous chapter. Similarly, let **SS4** [X; Y] and **SS5** [X; Y] be the logics obtained in a parallel fashion from **SS4** and **SS5**, respectively. In addition let **ST'** [X; Y] be **ST** [X + {T $\blacktriangle$ }; Y], and similarly for **SS4'** [X; Y] and **SS5'** [X; Y], where

T $\blacktriangle$ $T(\blacktriangle (A)) \leftrightarrow \blacktriangle (T(A))$

The above axiom is a reflection of the interaction of truth and modality. It appears to be intuitively sound for all the modals. Even in the case of belief it seems right: it is true that John believes that A iff John believes that A is true. Whatever the merits of these particular intuitions it is certainly sound for certain modals.

THEOREM 8.4

The logics **ST'** [X; Y], **SS4'** [X; Y] and **SS5'** [X; Y] are consistent logics of truth and predicative modality.

111

PROOF   Employ the translation. To establish the consistency of the logic **ST'**[X; Y] we employ the logic ST''[X; Y] from the previous chapter. We need only worry about the axioms of modality. For (r) we use the r axiom for the modal operator □ of the language *L3* followed by the R axiom for T. For (tr) observe that the following □ [T(A) → T(T(A))] is a theorem of the logic ST'[X; Y] of *L3*.

Assume □ (T(A)). By ip for □, □ [T(T(A))]. By tr for □, □ [□ (T(T(A)))]. By □ [□ (T(T(A))) ↔ T(□ (T(A)))] we obtain via, ip for □,

□ [T(□ (T(A)))], as required

T ▲ is left for the reader.

The logics generated by **SS4'**[X; Y] and **SS5'**[X; Y] are dealt with in a similar manner.   ●

In summary, we can *almost* obtain the standard logics but we must restrict the application of necessitation to proofs not involving any applications of IPT, r, R, tr, or Tr. We have throughout used the results and approach of 7.2. We leave the reader to reflect on the problems involved in employing the techniques of 7.1.

8.6   Logics of Truth, Propositions and Modality

The logics of this last section might be thought to be of little use given the restrictions on the employment of the necessitation rule. It is quite hard to work with a logic where one has to keep track of whether or not instances of certain axioms have been employed. Fortunately, there is a slightly different way of proceeding: we can impose restrictions on the rule itself rather than its application. To see how this might be feasible, we formulate a logic based upon a slight extension of the axioms of propositions and truth TP.

AXIOMS OF PROPOSITIONS

(i)   P(A) & P(B) → P(A & B)
(ii)   P(A) & P(B) → P(A ∨ B)
(iii)   P(A) & (T(A) → P(B)) → P(A → B)
(iv)   P(A) → P(~ A)
(v)   ∀xP(A) → P(∀xA)
(vi)   ∀xP(A) → P(∃xA)
(vii)   P(s = t)
(viii) P(A) → P(T(A))
(ix)   ▲ (P(A)) → P(▲ (A))

AXIOMS OF TRUTH

(i) $P(A) \& P(B) \rightarrow [T(A \& B) \leftrightarrow T(A) \& T(B)]$

(ii) $P(A) \& P(B) \rightarrow [T(A \lor B) \leftrightarrow T(A) \lor T(B)]$

(iii) $P(A) \& (T(A) \rightarrow P(B)) \rightarrow [T(A \rightarrow B) \leftrightarrow (T(A) \rightarrow T(B))]$

(iv) $P(A) \rightarrow [T(\sim A) \leftrightarrow \sim T(A)]$

(v) $\forall x P(A) \rightarrow [T(\forall x A) \leftrightarrow \forall x T(A)]$

(vi) $\forall x P(A) \rightarrow [T(\exists x A) \leftrightarrow \exists x T(A)]$

(vii) $T(s = t) \leftrightarrow s = t$

(viii) $T(A) \rightarrow P(A)$

(ix) $P(A) \rightarrow [T(T(A)) \leftrightarrow T(A)]$

(x) $\blacktriangle (P(A)) \rightarrow [T(\blacktriangle (A)) \leftrightarrow \blacktriangle (T(A))]$

The extensions concern the truth and modal predicates and are located in the last two axioms of each set. We have added the modal operator $\blacktriangle$ and the truth predicate T as first-class citizens of the theory. We leave the reader to add the clauses for $\blacktriangledown$. The logics we now entertain employ the notion of proposition to impose restrictions on the necessitation rule itself rather than its application. We add to the above logic of truth and propositions axioms and rules for modality.

Let **T'** be the logic with the above axioms for truth and propositions together with the following axioms and rules for modality:

$\blacktriangle (A) \rightarrow A$

$\blacktriangle (A \rightarrow B) \rightarrow (\blacktriangle (A) \rightarrow \blacktriangle (B))$

$\forall x \blacktriangle (A) \rightarrow \blacktriangle (\forall x A)$

If $\mathbf{T'} \vdash A \& P(A)$ then $\mathbf{T'} \vdash \blacktriangle (A)$

The important point is that the rule of necessitation is now restricted to propositions. We no longer have to keep track of which axioms have been employed. Instead we have to prove that the well-formed formulae in question is not only true but is also a proposition. This is to be established using the axioms of propositions.

Let **S4'** be $\mathbf{T'} + \{ \blacktriangle (A) \rightarrow \blacktriangle (\blacktriangle (A)) \}$. We have the following result.

THEOREM 8.5

T' and **S4'** are consistent logics of truth and predicative modality.

PROOF   We employ the translation and the logics $ST'[X; Y]$ of the last chapter. We really only need to check the soundness of the necessitation rule.

If $ST'[X; Y] \vdash A \& P(A)$ then $ST'[X; Y] \vdash T(A)$ and so $ST'[X; Y] \vdash \Box(T(A))$ as required. ●

In conclusion, the last result states that the modifications of the logics **T** and **S4**, where the necessitation rule is restricted to propositions, are available to us. So although we cannot have the full power of the standard modal systems for the logic of modality, we can have these restricted systems. Together with the axioms for propositions and truth given above, we have workable logics of syntactic modality. Finally, **S5′** is not available since the **S5** axiom for truth is not safe.

We have used **TP** for illustrative purposes but any of the logics from chapters 3 or 4 are possible candidates. Indeed, we have certainly not explored all the possible logics of truth and modality which are supported by the modal systems of chapter 7 but we trust that the reader is impressed by the number of possibilities that are available. The paradoxes do not prevent the development of interesting and useful logics of truth and modality. The richness of the predicative language (*L5*) has to be contained a little but there are several mechanisms available for achieving this. We hope that the reader will further explore the space of possibilities.

9 Conclusion

The logics of this book have all been introduced under the auspices of some semantic theory. In chapters 3, 4 and 5 we introduced three logics of truth by reflecting upon various semantic theories. In the second half of the book we invested these logics with further content by extending these theories to the modal context. At this point the semantic intuitions which underpin classical modal logic came into play. These, in combination with the semantic theories of truth, yielded different logics of truth and modality. Finally, in chapter 8 we employed these logics to develop logics of truth and predicative modality.

Our objective has not been to develop definitive logics of truth or modality but rather to draw boundaries around what is possible and map out the important features of the logical landscape. Nevertheless, it is desirable to briefly present the reader with some retrospective guide through the labyrinth of the logics available. We begin with the logics of truth.

9.1 Logics of Truth

Essentially, we have a choice between the fixpoint and stable theories. The former has the distinct advantage of simplicity. The various versions of the theory are easy to state axiomatically. Either the axioms based upon Kleene logic or the reformulation given in terms of the schemata C1 and C2 provide a logic of truth which is easy to apply. However, the theory has a certain conceptual drawback since it actually reflects Kleene's three-valued logic. As a consequence we have not formalized the classical concept of truth but rather the Kleene approximation to it. Of course, one could use a different three-valued (monotone) logic but this hardly addresses this criticism.

Inevitably, some principles of classical reasoning must be abandoned. In all the logics we have studied the principle of bivalence

$$T(B) \vee F(B)$$

is discarded. However, **KFG** goes further in that it abandons classical reasoning "internally". To understand what we mean by this, consider the following propositional logic of truth.

AXIOMS

P1 $T(A \to (B \to A))$

P2 $T(A \to (B \to C) \to ((A \to B) \to (A \to C)))$

P3 $T((\sim A \to B) \to ((\sim A \to \sim B) \to A))$

RULE

$$\frac{T(A) \qquad T(A \to B)}{T(B)}$$

The theorems of this logic are all the theorems of the form $T(A)$, where A is a theorem of classical propositional logic. The theory **KFG** does not support this logic. In particular, neither P1 not P2 are provable in **KFG**. It is in this sense that the logic **KFG** does not support classical reasoning internally.

Stable logic meets this desideratum since the model-theoretic process of revision is based firmly upon the classical truth conditions. As a consequence, both the above logic and its Predicate Calculus generalization are sound under the stability interpretation. The disadvantage of this approach concerns the number of logics available; there is no simple characterization of stable truth as there is with the fixpoint theories. However, the logics of stable truth take on a very systematic pattern. The logics **ST**, **SS4** and **SS5** closely parallel the logics **T**, **S4** and **S5** of standard modal logic. These logics are easy to use since full necessitation is sound and consequently they possess the laudable property of rendering all logical truths *true*. Furthermore, they appear to possess a certain intrinsic mathematical interest.

This leaves us to deal with the logic **TP** which underlies Frege Structures. This is a simple and elegant logic and seems to be the minimum required of a theory of truth. Indeed, in applications to knowledge representation it may be sufficient, but this remains to be seen.

9.2 Modality

Modality introduces a new dimension of choice. The different modal logics combine with the various logics of truth to yield a rich variety of possible logics of truth and modality. If we are right about the theory **TP**, then the logics **T'** and **S4'** of chapter 8 should prove sufficient for most applications. The dimension of choice is also more complex in the modal setting since the choice of the modal logic will be determined by other factors. The particular modality under scrutiny will play a major role. Logics of knowledge and belief are, conceptually, notoriously difficult. Our task, however, has not been to argue for any particular logic of modality but rather to draw a circle around what is possible.

This rich variety of choice and the corresponding complexity of such mixed modal logics is a common feature of much formal work in knowledge representation. Such formal systems have to reflect a wide variety of common sense notions and their formal relationships. Undoubtedly, the study and development of such systems will add a new dimension to philosophical logic. Not only is there an exponential growth of possibilities as the number of different modal notions increases, but the interaction of the various notions raises new conceptual and mathematical questions which are only pinpointed by the particular combination of concepts under study.

References

Aczel, P. [1980] "Frege structures and the notions of proposition, truth and set" in *The Kleene Symposium* (Eds. Barwise, Keisler, Keenan), North Holland Studies in Logic, pp. 31–39.

Allen, J. F. [1984] "Towards a general theory of action and time", *Artificial Intelligence*, vol. 23 (2), pp. 123–154.

Bealer, G. [1982] *Quality and Concept*, Clarendon Press, Oxford.

Barendregt, H. [1984] *The Lambda Calculus: its Syntax and Semantics*, North Holland Studies in Logic and the Foundations of Mathematics, vol. 103.

Bochvar, D. A. [1939] "On a three-valued logical calculus and its application to the analysis of contradictories", *Matèmatičéskij Sbornik* (4), 1939.

Boolos, G. [1979] *The Unprovability of Consistency*, Cambridge, University Press.

Dowty, D., Wall, R. and Peters, S. [1981] *Introduction to Montague Semantics*, D. Reidel..

Feferman, S. [1979] "Constructive theories of functions and classes" in *Logic Colloquium* (Eds. Baffa, Van-Dolen, McAloon), North Holland Studies in Logic and the Foundation of Mathematics, vol. 78, pp. 159–224.

Feferman, S. [1984] "Towards useful type-free theories 1", *Journal of Symbolic Logic*, vol. 49, pp. 75–111.

Gilmore, P. C. [1974] "The consistency of partial set theory without extensionality", *Axiomatic Set Theory*, Proc. Symposia Pure Maths, vol. XIII, Part II, pp. 147–153, Amer. Math. Soc.

Gupta, A. [1982] "Truth and paradox", *Journal of Philosophical Logic*, vol. 11, pp. 1–60.

Herzberger, H. [1982] "Notes on naive semantics", *Journal of Philosophical Logic*, vol. 11, pp. 61–102.

Hindley, R. and Seldin, J. [1986] *Introduction to Combinators and the Lambda Calculus*, London Mathematical Society Student Texts 1, Cambridge University Press.

Hintikka, J. [1962] *Knowledge and Belief: an Introduction to the Two Notions*, Cornell University Press.

Hughes, G. and Cresswell, M. [1968] *Introduction to Modal Logic*, Methuen, London.

Hughes, G. and Cresswell, M. [1984] *A Companion to Modal Logic*, Methuen, London.

Kamp, H. and Ascher, N. [1988] "Self-reference, attitudes and paradox" in *Properties, Types and Meaning* (Eds. Chierchia, G., Partee, B. & Turner, R.), vol. 1, Kluwer Academic Press.

Kleene, S. C. [1967] *Mathematical Logic*, Wiley, London.

Konolige, K. [1986] *A Deduction Model of Belief* (Research Notes in Artificial Intelligence Series), Pitman, London.

Kripke, S. [1963] "Semantical considerations on modal logic", *Acta Philosophica Fennica*, vol. 16, pp. 83–89.

Kripke, S. [1975] "Outline of a theory of truth", *Journal of Philosphy*, vol. 1, xxii, pp. 690–716.

McDermott, D. V. [1982] "A temporal logic for reasoning about processes and plans", *Cognitive Science*, vol. 6, pp. 101–155.

Martin-Löf, P. [1979] "Constructive mathematics and computer programming" in *Logic, Methodology and Philosophy of Science VI* (Eds. Cohen, Los, Pfeiffer and Podewski), pp. 153–179, North Holland 1982.

Montague, R. [1963] "Syntactical treatment of modalities, with corollaries on reflexion principles and finite axiomatizability", *Acta Philosophica Fennica*, vol. 16, pp. 153–167.

Montague, R. [1973] "The proper treatment of quantification in ordinary English" in *Approaches to Natural Language* (Eds. Hintikka, Moravcsik and Suppes), Dordrecht 1973.

Montague, R. and Kaplan, D. [1960] "A paradox regained", *Notre Dame Journal of Formal Logic*, vol. 1, pp. 79–90.

Ramsay, A. [1988] *Formal Methods in Artificial Intelligence*, Cambridge Texts in Theoretical Computer Science 6.

Perlis, D. [1986] "Languages with self-reference I", *Artificial Intelligence*, vol. 25, pp. 301–322.

Perlis, D. [1988] "Languages with self-reference II: knowledge, belief and modality", *Artificial Intelligence*, vol. 34, pp. 179–212.

Sainsbury, M. [1988] *Paradoxes*, Cambridge University Press.

Scott, D. [1973] "Models for various type-free calculi" in *Logic, Methodology and Philosophy of Science IV* (Ed. P. Suppes *et al*), North Holland Studies in Logic and the Foundations of Mathematics, pp. 157–187.

Scott, D. [1975] "Combinators and classes" in *Lambda Calculus and Computer Science Theory*, Lecture Notes in Computer Science, vol. 37, Springer-Verlag, Berlin.

Smorynski, C. [1985] *Self-Reference and Modal Logic*, Springer-Verlag.

Tarski, A. [1937] "The concept of truth in formalized languages" in *Logic, Semantics and Metamathematics*, pp. 152–278, Clarendon Press.

Thomason, R. [1980] "A model theory for the propositional attitudes", *Linguistics and Philosophy*, vol. 4, pp. 47–70.

Thomason, R. [1986] "Paradoxes and semantic representation" in *Reasoning about Knowledge* (Ed. Halpern), pp. 225–239, Morgan Kaufmann Los Altos, Cal.

Turner, R. [1987] "A theory of properties", *Journal of Symbolic Logic*, vol. 52, no. 2, pp. 445–472.

Turner, R. [1988] "Two issues in the foundations of semantic theory" in *Properties, Types and Meaning*, vol. 1.

Turner, R. [1989] "Properties, propositions and semantic theory" in *Proc. of Formal Semantics and Computational Linguistics*, Lugano, Switzerland, 1988.

Visser, A. [1984] "Four-valued semantics and the liar", *Journal of Philosophical Logic*, vol. 13, pp. 181–212.

Woodruff, R. L. [1975] "On representing 'True-in-L'" in *Philosophia*, vol. 5, pp. 213–217.

Formal Terms Index

General Index